AF614792

End-of-Life Care in Cardiovascular Disease

Sarah J. Goodlin • Michael W. Rich
Editors

End-of-Life Care in Cardiovascular Disease

Editors
Sarah J. Goodlin, MD
Patient-Centered Education and Research
Salt Lake City, Utah
USA

Portland VA Medical Center
Portland, OR
USA

Michael W. Rich, MD
Washington University School of Medicine
St Louis, Missouri
USA

ISBN 978-1-4471-6520-0 ISBN 978-1-4471-6521-7 (eBook)
DOI 10.1007/978-1-4471-6521-7
Springer London Heidelberg New York Dordrecht

Library of Congress Control Number: 2014950791

Printed on acid-free paper

Springer is part of Springer Science+Business Media (www.springer.com)

Foreword

I have heard so many bad stories of life at its end with heart disease, and so many of them were made distressing and frustrating by the actions of the health care team that just meant to do good for the patient. There's the poor fellow dying of pancreatic cancer whose implanted defibrillator fired off twenty times in as many hours while he debated with himself about the moral status of turning it off – having never considered the issue until his own last hours. There's the person drowning in fluid begging in gasps to make this stop, but no one was willing and able to provide sedation when the only other option was to die of suffocation and exhaustion. And there's the oft-repeated story of the bereaved family that did not get to say good-byes, because they never really knew that Dad's heart trouble would be fatal, or that death would likely come abruptly.

We don't so often hear of the people who live reasonably rewarding and comfortable lives to the end, whose dying was peaceable and meaningful to family and friends, and whose story, when told, ends with "Weren't we lucky!" In a sense, this collection of information and wisdom is meant to make it unnecessary to be lucky in order to get good care at the end of a life that is lived with heart disease. Instead, the authors aim to build understandings, processes, and expectations that enable heart patients to live well as they live in the shadow of a fatal illness. These chapters provide concrete protocols where that is what matters, remind us to think ahead and make plans while the patient still can, and urge us to pay attention to how well we are doing.

To live long with illnesses that once took life suddenly and often at an early age – that's a major advance in medical care, and patients and families are glad to have the time that has been gained. The price of that success should not be misery at life's end. It is time to help clinicians ensure that their heart disease clients live well right to the end of life, and these writings provide the guidance needed.

Joanne Lynn, MD, MA, MS　　　　Washington, DC, USA

Preface

Over the past 50 years, there have been remarkable advancements in medical care and interventions to prevent and treat heart disease, and death rates from cardiac disease have declined correspondingly. For example, fewer patients die from acute myocardial infarction; rather, most patients survive the initial event but die, often years later, from heart failure or other illnesses. Similarly, patients with life-threatening valvular heart disease, congenital heart abnormalities, and heart failure live longer with fewer symptoms as a result of medical, percutaneous, and surgical interventions. Thus, while still life-shortening, cardiovascular diseases have become primarily chronic illnesses. Persons with cardiovascular disease are increasingly living into old age, and in the process, they are accumulating other chronic conditions that adversely impact the quality of life in multiple dimensions. Yet, a substantial proportion of these individuals ultimately die from cardiovascular causes, and cardiovascular disease remains the leading cause of death worldwide in both men and women. Moreover, as the world population ages, the global burden of cardiovascular disease will continue to rise [1].

In developed countries, the majority of patients dying with cardiovascular disease are elderly and have significant comorbidities. Indeed, elderly persons with heart failure, coronary disease, or cerebrovascular disease almost invariably have other illnesses unrelated to their cardiovascular disease, and they often die from these other conditions. For example, of the nearly 300,000 people in the United States who die each year with a mention of heart failure on the death certification, heart failure is listed as the primary cause of death for only about 20 % [2].

Despite the high probability of death from cardiovascular causes in patients with prevalent cardiovascular disease including stroke, clinicians often fail to discuss prognosis and end-of-life preferences with patients and families. Conversely, patients themselves, even when they recognize failing health, are reluctant to bring their concerns to the attention of their doctor or other health-care provider. This reluctance on the part of both clinicians and patients in part reflects the greater difficulty in defining disease trajectory and prognosis in patients with cardiovascular disease compared to those with advanced cancer or HIV/AIDS. Thus, the approach to the end of life in cardiovascular disease is not only less acknowledged but also less well understood and more difficult to recognize. These gaps in knowledge and skills, in end-of-life care for cardiovascular illness frequently present challenges in clinical practice.

This book, *End-of-Life Care in Cardiovascular Disease*, is designed to assist clinicians, nurses, and other health-care providers in addressing end-of-life care for patients with cardiovascular disease in a variety of common clinical scenarios. Each chapter is written by expert clinicians and researchers, and we are grateful for their hard work, which is evident in the exceptional quality of their contributions.

While palliative and supportive care should be provided throughout the course of serious and life-shortening illnesses, we focused on issues that clinicians encounter in end-of-life care. Some problems such as frailty and cognitive impairment add significant complexity to the management of persons with cardiovascular disease especially at the end of life. Devices and their management present unique challenges, as do interventions in the intensive care unit, and we intend this book to provide guidance to clinicians working with advanced technologies.

The epidemiology and tools to identify patients near the end of life are addressed in Chap. 1. The book's scope includes specific sites of care:

- Acute hospital
- Intensive care unit
- Emergency room
- Skilled nursing facility
- Home

as well as special populations of patients with cardiovascular disease:

- Pediatric patients and those with congenital heart disease
- Patients with advanced heart failure
- Patients with arrhythmias

Several fundamental issues are important to end-of-life care in all settings:

- Communication and decision-making about alternatives in care, particularly in an era of patient-centered care
- Symptom assessment and management
- Spirituality and bereavement assessment and support

These fundamentals are increasingly important to clinicians caring for patients who die an unexpected or sudden death in the community, in the emergency department, or during an acute hospitalization. We intend this book to be a resource to these clinicians as well.

Creation of this book brought attention to the need to increase knowledge about the topics addressed. While the end of life is only one phase in the care of patients with significant illness, we must strive to "get it right." In particular, there is a compelling need for research that provides evidence to support the best care for patients' symptoms, including what therapies and interventions are most effective for which patients. We need to understand how to enhance clinicians' communication and decision-making skills, as well as how to make communication culturally and ethnically sensitive and patient and family centered. Lastly as clinicians, we need to recognize our own feelings of sadness associated with clinical decline and death and how to best support the interdisciplinary groups of providers who care for patients dying with cardiovascular disease as well as patients and their families.

We hope that readers of this book will find it to be a valuable resource in caring for patients with cardiovascular disease at the end of life. We welcome any comments or suggestions you may have.

References

1. Moran AE, Forouzanfar MH, Roth GA et al. Temporal trends in ischemic heart disease mortality in 21 world regions, 1980–2010: the global burden of disease 2010 study. Circulation. 2014;129:1483–92. doi: 10.1161/CIRCULATIONAHA.113.004042.
2. Go AS, Mozaffarian D, Roger VL et al. Heart disease and stroke statistics–2014 update: a report from the American Heart Association. Circulation. 2014;129(3):e28–e292. doi: 10.1161/01.cir.0000441139.02102.80.

Sarah J. Goodlin, MD — Portland, OR, USA
Michael W. Rich, MD — St. Louis, MO, USA

Acknowledgment

The editors would like to thank Ms. Connie Walsh for her outstanding work as developmental editor for this monograph and for her invaluable assistance in shepherding it through to completion.

Contents

Contributors

Husam Abdel-Qadir, MD Department of Medicine, Toronto General Hospital, Toronto, ON, Canada

Jonathan Afilalo, MD, MSc, FACC, FRCPC Division of Cardiology, Jewish General Hospital, McGill University, Montreal, QC, Canada

David B. Bekelman, MD, MPH Department of Internal Medicine, VA Eastern Colorado Healthcare System and the University of Colorado School of Medicine, Denver, CO, USA

Richard D. Brumley, MD Hospice and Palliative Medicine, Laguna Niguel, CA, USA

Christopher R. Carpenter, MD, MSc Department of Emergency Medicine, Washington University in St. Louis School of Medicine and Barnes-Jewish Hospital, St. Louis, MO, USA

Maria Dans, MD Department of Medicine, Washington University School of Medicine, St. Louis, MO, USA

Palliative Care Service, Internal Medicine, Barnes-Jewish Hospital, St. Louis, MO, USA

Michael G. Dickinson, MD Richard DeVos Heart & Lung Transplant Program, Frederik Meijer Heart and Vascular Institute, Spectrum Health, Grand Rapids, MI, USA

Michigan State University, Grand Rapids, MI, USA

Pablo Díez-Villanueva, MD, PhD Department of Cardiology, Hospital Universitario Gregorio Marñón, Madrid, Spain

Anne I. Dipchand, MD Heart Transplant Program, Division of Cardiology, Hospital for Sick Children, Toronto, ON, Canada

Susan Enguidanos, MPH, PhD Leonard Davis School of Gerontology, University of Southern California, Los Angeles, CA, USA

Kathleen Garcia, MD Department of Medicine, Barnes-Jewish Hospital, St. Louis, MO, USA

Matthias Greutmann, MD Adult Congenital Cardiology Program, Cardiology Department, University Hospital Zurich, Zurich, Switzerland

Adam Herman, BSc Division of Cardiology, Department of Medicine, Toronto General Hospital, University Health Network, Toronto, ON, Canada

Stephanie A. Hooker, MS Department of Psychology, University of Colorado Denver, Denver, CO, USA

Aluko A. Hope, MD, MSCE Division of Critical Care Medicine, Department of Medicine, Montefiore Medical Center, Bronx, NY, USA

Susan M. Joseph, MD Cardiovascular Division, Department of Internal Medicine, Washington University in St Louis, St. Louis, MO, USA

Corrine Y. Jurgens, PhD, RN, ANP-BC, FAHA School of Nursing, Stony Brook University, Stony Brook, NY, USA

Adrienne H. Kovacs, PhD, CPsych Toronto Congenital Cardiac Centre for Adults, Peter Munk Cardiac Centre, University Health Network, Toronto, ON, Canada

Rachel Lampert, MD Section of Cardiology, Yale University School of Medicine, New Haven, CT, USA

Douglas S. Lee, MD, PhD Department of Medicine, Institute for Clinical Evaluative Sciences, University of Toronto, Toronto, ON, Canada

Division of Cardiology, Department of Medicine Toronto General Hospital, University Health Network, Toronto, ON, Canada

Hannah I. Lipman, MD, MS Divisions of Geriatrics and Cardiology, Medicineand The Montefi ore Einstein Center for Bioethics, Montefi ore Medical Center, Bronx, NY, USA

Derrick Lowery, MD Hospice and Palliative Medicine Fellow, Department of Internal Medicine, Division of Geriatric Medicine and Gerontology, Emory University, Atlanta, GA, USA

Joanne Lynn, MD, MA, MS Center on Elder Care and Advanced Illness, Altarum Institute, Washington DC, USA

Daniel L. Maison, MD Palliative Care, Spectrum Health Medical Group, Spectrum Health, Grand Rapids, MI, USA

Manuel Martínez-Sellés, MD, PhD Cardiology Department, Hospital Universitario Gregorio Marñón and Universidad Europea, Madrid, Spain

Christopher M. Meeusen, MD Internal Medicine Department, Spectrum Health Butterworth, Grand Rapids, MI, USA

Caroline Michel, MD, FRCPC Division of Cardiology, Jewish General Hospital, McGill University, Montreal, QC, Canada

Diane K. Pastor, PhD, MBA, NP-C Adult Health Program, School of Nursing, Stony Brook University, Stony Brook, NY, USA

Craig Tanner, MD General Medicine and Geriatrics, Portland VA Medical Center, Portland, OR, USA

Medicine Department, Oregon Health and Science University, Portland, OR, USA

Daniel Tobler, MD Cardiology Department, University Hospital Basel, Basel, Switzerland

Justin M. Vader, MD Cardiovascular Division, Department of Internal Medicine, Washington University in St Louis, St. Louis, MO, USA

Chapter 1
Dying from Cardiovascular Disease: An Epidemiologic Perspective

Husam Abdel-Qadir, Adam Herman, and Douglas S. Lee

Abstract While cardiovascular disease is widely recognized as a leading cause of death, end of life care for patients dying from cardiovascular disease remains incompletely embraced. Heart failure is a common final pathway for many such patients; its incidence is increasing as are the costs associated with it. A significant proportion of the costs associated with congestive heart failure are incurred in the last months of life, and may be related to overly aggressive care in patients with an irreversibly poor prognosis. Many factors contribute to the poor uptake of palliative and hospice care. Chief amongst these is the difficulty in establishing prognosis and predicting death among patients with cardiovascular disease. It has been suggested that novel models of palliation may be needed given these challenges.

In this chapter, we discuss the epidemiology of dying from heart disease, and highlight some of the expected benefits with palliative/hospice care while highlighting gaps in its provision and uptake. We provide an overview of practical, validated methods for predicting prognosis near end of life for patients with cardiovascular illness that are suitable for use at the bedside. We also provide practical guidance to facilitate use of clinical epidemiological principles for patient management and resource use near end of life is provided.

Keywords Death • Palliative care • Prediction • Prognosis • Epidemiology • Cardiovascular disease • Heart failure • Utilization

---

H. Abdel-Qadir, MD
Department of Medicine, Toronto General Hospital, Toronto, ON, Canada

A. Herman, BSc
Division of Cardiology, Department of Medicine, Toronto General Hospital, University Health Network, Toronto, ON, Canada

D.S. Lee, MD, PhD (✉)
Department of Medicine, Institute for Clinical Evaluative Sciences, University of Toronto, Toronto, ON, Canada

Division of Cardiology, Department of Medicine, Toronto General Hospital, University Health Network, NU 4-482 200 Elizabeth St., Toronto, ON M5G 2C4, Canada
e-mail: dlee@ices.on.ca

S.J. Goodlin, M.W. Rich (eds.), *End-of-Life Care in Cardiovascular Disease*,
DOI 10.1007/978-1-4471-6521-7_1

Key Points

- As a leading cause of death and disability, cardiovascular diseases are one of the major areas where palliative care is needed.
- Congestive heart failure is a common terminal pathway for many cardiovascular illnesses, and has been the best studied form of cardiovascular disease near end of life.
- Although palliative care is beneficial for patients dying with cardiovascular disease, it remains underused within this population in part due to the difficulty in estimating end-of-life prognosis.
- Multiple prognostic tools have been developed to estimate prognosis in patients with heart failure and other cardiovascular illness, and they can serve as valuable tools in supplementing clinical judgment to select which patients might benefit from palliative care.
- Current models of palliative care are derived from those developed for patients with cancer, and may be suboptimal for patients dying from cardiovascular illness. Newer models, emphasizing care at home, may be needed for this patient group.

Introduction: Impact of End of Life Care in Cardiovascular Disease

Globally, there are 55 million deaths per year, and it is estimated that approximately 300 million patients and family members are affected by medical end of life issues annually [1, 2]. Non-communicable diseases account for two-thirds of all deaths globally, of which cardiovascular disease is the leading cause accounting for approximately 30 % of all deaths world-wide [1]. As a consequence, cardiovascular diseases are the major condition necessitating palliative care globally [3–5]. Approximately one in three persons have some form of cardiovascular disease, and it is estimated that nearly 45 million Americans will be projected to have cardiovascular disease by the year 2030 [6]. Longitudinally, however, the overall number of deaths from cardiovascular disease has decreased between 2004 and 2011. During this period, the proportion of all cardiovascular deaths due to acute coronary syndromes have declined, while chronic coronary heart disease deaths have not decreased [7].

Congestive heart failure is a common final pathway for most forms of cardiac disease and is associated with a grim prognosis. One in nine death certificates cite heart failure as a causal or contributing factor [6]. After an initial hospitalization for congestive heart failure, the 10-year mortality rate has been reported to be 98.8 %, with an average of four hospital readmissions per patient occurring over the lifetime of a well-studied cohort [8]. A sizable proportion of these hospitalizations occur near the end of life, with an estimated 40 % of heart failure and 52 %

of all cardiovascular readmissions occurring in the last decile of the cohort survival duration [8]. Heart failure is most often considered the transitional state from most other forms of cardiovascular disease to an end of life trajectory, and consequently, it is the primary cardiac condition that has been associated with palliative approaches to care.

Patients diagnosed with heart failure in the community also exhibit poor long-term prognosis, although to a lesser extent than hospitalization-based cohorts. Data from the Framingham Heart Study indicated that the 30-day, 1-year, and 5-year age-adjusted mortality rates between 1990 and 1999 were 11, 28, and 59 %, respectively among men [9]. The corresponding rates were 10, 24, and 45 % among women [9]. Another study from Olmstead County, Minnesota, also reported an age-adjusted 5-year mortality estimate of 48 % between 1996 and 2000 [10]. The trends in heart failure mortality are likely to be influenced by temporal changes in the subtypes of heart failure. The incidence of heart failure with preserved ejection fraction (HFpEF) is likely to rise in the future as the population ages [11–13]. However, the mortality rate of patients with HFpEF has remained static while outcomes for heart failure with reduced ejection fraction (HFrEF) have improved due to the availability of therapies that improve survival for the latter group [11].

Congestive heart failure increases in prevalence with increasing age. Thus, these patients often have a high burden of comorbidities that contribute towards their high mortality rates. This is particularly notable in patients diagnosed with HFpEF. An examination of the decedents of the Framingham Heart Study with a history of heart failure revealed that 70 % of patients with HFrEF, and only 44 % of those with HFpEF died of a cardiovascular cause (see Fig. 1.1) [14]. Thus, a reduced ejection fraction is predictive of death from a cardiovascular cause among patients with heart failure, with odds ratios of 3.16 (95 % confidence interval [CI], 1.73–5.78) in men and 2.39 (95 % CI, 1.39–4.08) in women. Among patients with HFrEF who died of a cardiovascular cause, 66 % died from progressive pump failure while 32 % died from arrhythmias or sudden cardiac death. Among patients with HFpEF, the most common cardiovascular causes of death were circulatory failure/heart failure occurring in 50 % and arrhythmia/sudden cardiac death occurring in 27 %. Infections and kidney disease were identified as the key immediate and contributing causes of death, respectively in patients dying from non-cardiovascular causes. Prior myocardial infarction was associated with increased risk of cardiovascular death in women with HF with odds ratio 1.87 (95 % CI, 1.10–3.16) but not in men. Increasing age also decreased the likelihood of a cardiovascular cause of death in this cohort [14].

A similar analysis of decedents with heart failure from Olmstead County reported that 43 % of deaths were non-cardiovascular and preserved ejection fraction was associated with a marginally lower risk of cardiovascular death [15]. In contrast, the leading cause of death in subjects with HFrEF was coronary heart disease (43 %). The proportion of cardiovascular deaths decreased significantly from 69 in 1979–1984 to 40 % in 1997–2002 among subjects with preserved left ventricular ejection fraction, in contrast to a more modest change among those with HFrEF from 77 to 64 % [15].

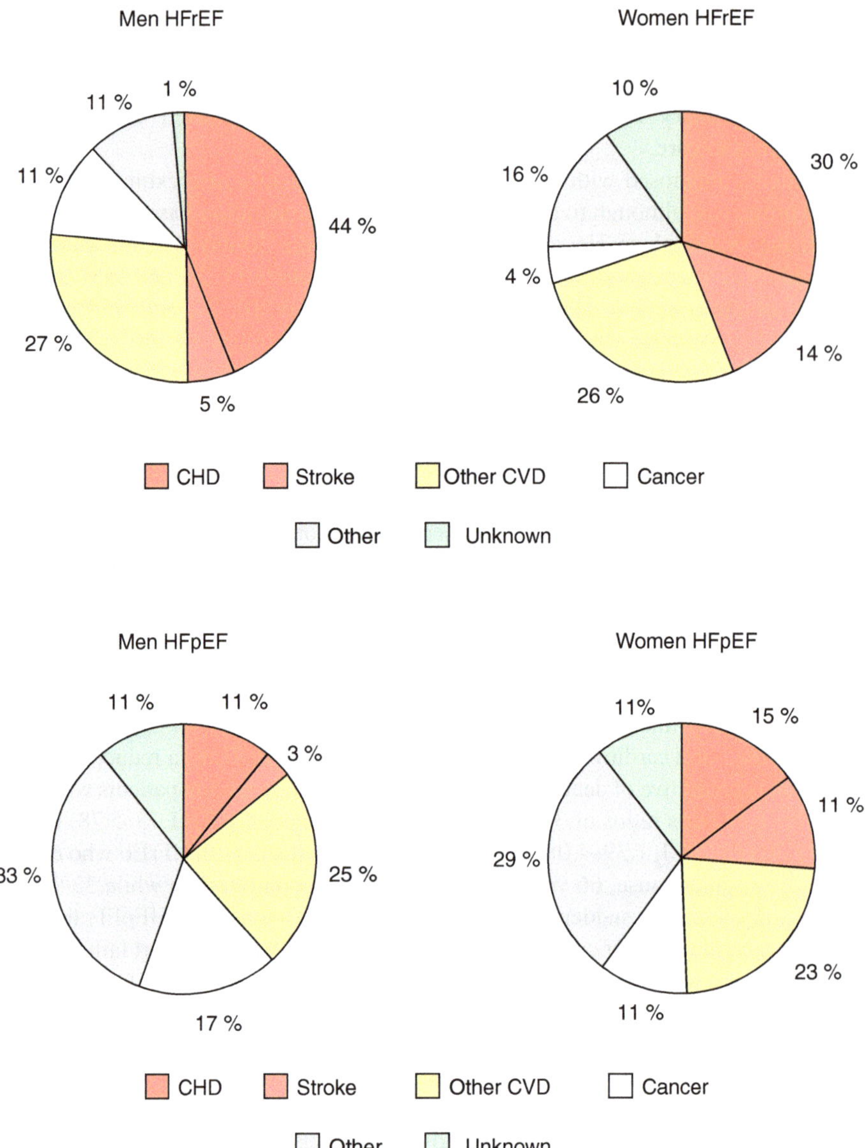

Fig. 1.1 Causes of death in men and women with HFrEF vs. HFpEF (From Lee et al. [14]. Reprinted with permission from Wolters Kluwer Health)

Palliative Care in Cardiovascular Illness

Patients dying with cardiovascular disease exhibit both high rates of hospital use and geographic variations in care patterns. In a retrospective analysis of Medicare beneficiaries with heart failure who died, 80 % were hospitalized within the last 6 months of life, although the majority of hospitalizations were for diagnoses other than heart failure [16]. Over 8 years, the mean length of stay in hospital remained constant, yet the average length of stay in intensive care increased from 3.5 to 4.6 days and the percentage of patients discharged from hospital to hospice care rose. Accordingly, the costs incurred to Medicare by patients with heart failure near the end of life increased by 11 % between 2000 and 2007 after adjustment for age, sex, race, comorbid conditions, and geographic region [16]. It has been estimated that the average cost of heart failure care in the last 6 months of life is more than five-fold the per-capita health care expenditure [17]. The staggering impact of these large per patient numbers is compounded by the expected rise in the prevalence of heart failure as a population with a high burden of cardiovascular disease and its risk factors continues to age [18].

Over the past decade, there has been increasing awareness of the need to include palliative care in heart failure treatment plans. In a 2004 survey of clinically active members of the Heart Failure Society of America, two-thirds of respondents had not referred a single patient for palliation in the preceding 6 months and almost 88 % had referred less than six patients over that time period [19]. While this survey was limited by a poor response rate of 24 %, such results support the notion that "cardiologists often focus on what can be done rather than what should be done, and the latter consideration may be neglected in the midst of therapeutic optimism" [20]. There has been little change in the utilization of costly invasive cardiac procedures in the last 6 months of life, including cardiac catheterization, pacemaker or implantable cardioverter-defibrillator implantation, and coronary artery bypass graft surgery, while the use of echocardiography has increased within the last 6 months of life of patients with heart failure [16].

The utilization of palliative care or hospice services is relatively low among patients with cardiovascular disease relative to other terminal illnesses such as cancer. In one population of medically insured patients, only 20 % of patients with end-stage heart failure were referred to hospice compared to 51 % of cancer patients. Opiate prescriptions have been proposed as a surrogate for palliative treatments and were observed to be used in 22 % of heart failure compared with 46 % of cancer patients [21]. Temporally, there has been only modest uptake of hospice and/or palliative care among patients with end-stage heart failure [21–23], despite the class I recommendation that palliative/hospice care referral should be offered to patients with end-stage heart failure in the American College of Cardiology/American Heart Association [24] and other international guidelines [25, 26]. Certain characteristics appear to predict a greater likelihood of referral to hospice or palliation in advanced heart failure: younger age, male gender, white ethnicity, higher income, and dialysis dependence [21–23, 27]. In addition, hospitals in greater compliance with heart failure performance measures may be more likely to refer patients with advanced heart failure to hospice [22].

There are multiple potential explanations for the persistently aggressive care and low rates of utilization of palliative care and hospice services among patients with heart failure and other cardiovascular diseases:

- Palliation is not traditionally part of the 'therapeutic culture' of cardiology.
- Palliative care is closely aligned with oncology and in some countries funding is aligned with cancer services, leading to non-acceptance of cardiac patients for palliative care.

- There is a paucity of evidence-based palliative care of patients with cardiovascular disease.
- A significant proportion of patients succumb to sudden cardiac death thus limiting the opportunity for provision of palliative care.
- The most commonly-cited barrier to hospice referral and end of life care in patients with cardiovascular disease is the unpredictable nature of the illness, and difficulty in determining when death will occur [28–31].

While cardiac specialists are familiar with the cardinal features of cardiovascular disease, patients may develop a multitude of other symptoms that markedly decrease their quality of life. Comparisons of outpatients with heart failure and advanced cancer reveal an equivalent burden of physical and depressive symptoms; patients with worse heart failure functional class had greater physical symptom burden, higher depression scores, and lower spiritual well-being than patients with advanced cancer [32]. As shown in Fig. 1.2, heart failure patients experience numerous symptoms, which are not part of the usual cardiovascular health evaluation. Moreover, the symptom burden in heart failure patients is an important prognostic factor. In the Carvedilol or Metoprolol European Trial (COMET), the symptom of breathlessness remained significantly associated with death while fatigue was the primary predictor of worsening heart failure in multivariate analysis [34]. Moreover, anxiety and depression have been consistently linked to worse outcomes in cardiovascular illness, with some of the postulated mechanisms involving detrimental alterations in physiology [35, 36].

Physicians and nurses often incompletely assess the symptom burden outside the cardinal manifestations of cardiovascular disease [37, 38]. There is also a significant gap between physician and patient perspectives on the degree of functional limitation, as shown in Fig. 1.3. Commonly used scales such as the NYHA functional classification do not account for other limiting symptoms such as pain, depression, or nausea. Tools that provide a global assessment of symptom burden to allow for effective management include the Memorial Symptom Assessment Scale [40], Edmonton Symptom Assessment Scale [41, 42], Quality-of-Life at the End of Life [43], the McGill Quality of Life Questionnaire [44], and the Palliative Performance Scale [45].

Models of Care at the End of Life

At end of life, patients receiving palliative care services often also receive improved holistic care, continuity of care, more focused goals of care, and attain better reported measures of anxiety, depression, global health status, and physical, social, cognitive, and emotional function [46, 47]. Paradoxically, observational studies have shown longer survival in patients referred for hospice care [48]. However, HF patients' multiple active cardiac issues may be best managed by close collaboration between the different disciplines [49, 50]. Primary health care providers shoulder a significant burden of heart failure patients' care to facilitate home deaths [51].

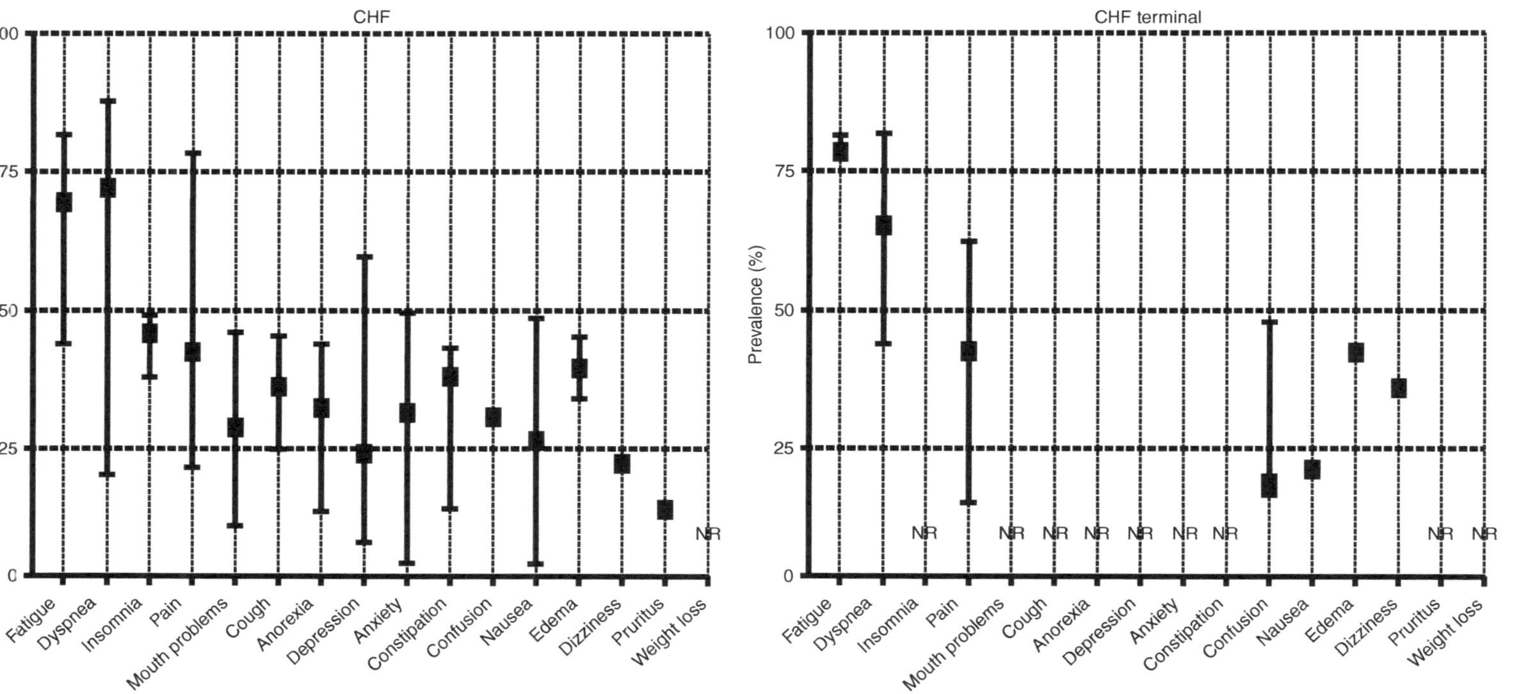

Fig. 1.2 Minimum, maximum and median prevalence of reported daily symptom burden in patients with end-stage congestive heart failure (*CHF*, *left upper panel*). *NR* not reported (From Janssen et al. [33]. Reprinted with permission from SAGE Publications)

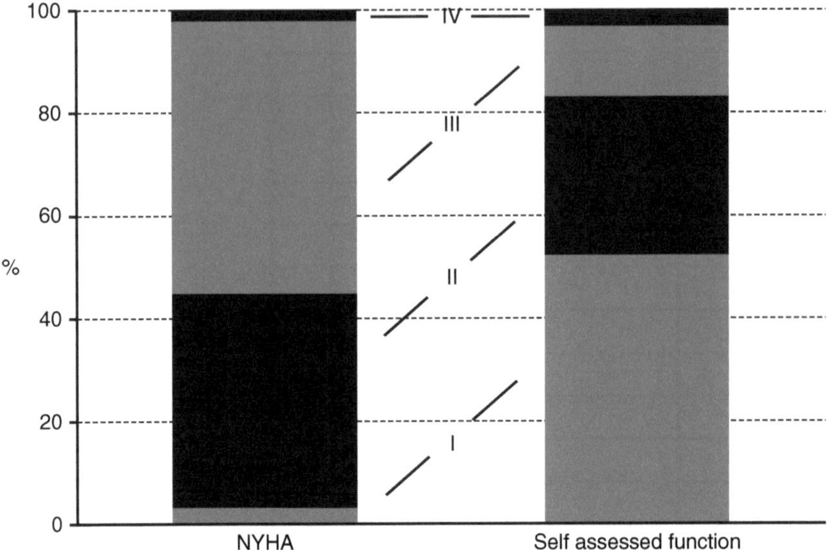

Fig. 1.3 Patients' self-assessed functional classification compared to conventional NYHA classification (From Ekman and Cleland [39]. Reprinted with permission from Oxford University Press)

However, the impact of hospice referrals on costs is unclear, with conflicting reports from different studies [52–54]. While rates of hospitalization, days in intensive care, and invasive procedures are reduced, the use of hospice increased overall Medicare expenditures compared to non-hospice care after 154 days for approximately 15 % of HF patients [54]. A slightly larger percentage of patients with heart failure referred for hospice care are alive 6 months after referral than patients with cancer [54, 55]. In fact 19 % of patients with heart failure are discharged alive from hospice compared with 11 % of those with cancer diagnosis [55]. Alternate models including outpatient palliative care may be needed for patients with cardiovascular disease [47, 55–60].

Estimating Survival Among Patients with Cardiovascular Disease Near End of Life

One of the major reasons for the difficulty in prognosticating end of life is because of the stochastic nature to cardiovascular deaths. Changes in functional status are often unpredictable and not closely linked with imminent death [61, 62]. Patients also have misperceptions about their prognosis when compared to estimates of their life expectancy based on multivariate models [62, 63]. The trajectory of functional decline in patients dying with heart failure is often different from cancer, being characterized by episodes of acute worsening followed by an often rapidly

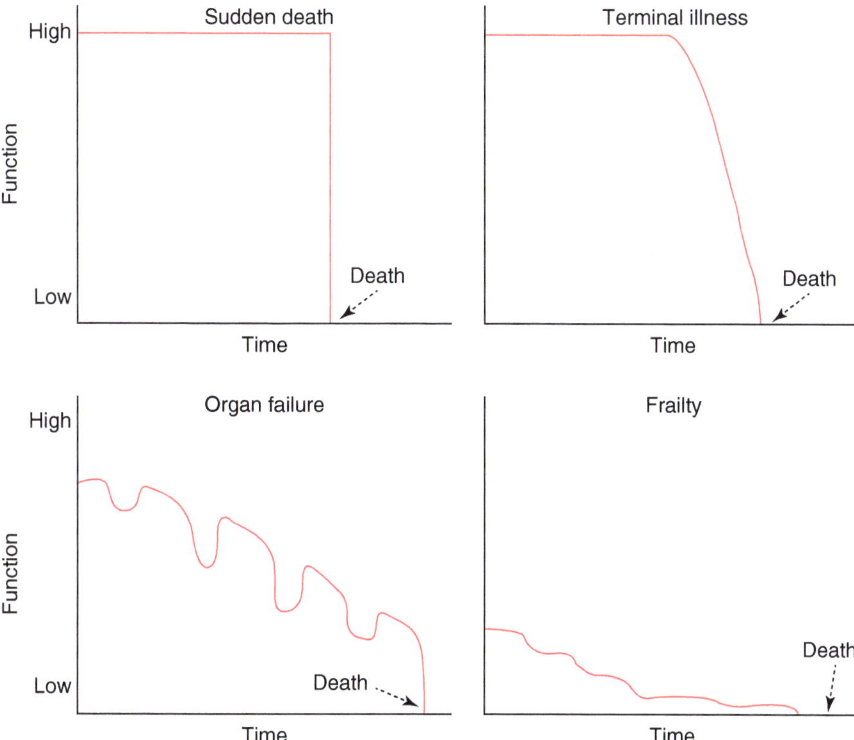

Fig. 1.4 Theoretical trajectories of death (From Lunney et al. [65]. Reprinted with permission from John Wiley and Sons)

progressive terminal phase [64]. Of four hypothetical trajectories near end of life: sudden death, terminal illness, organ failure, and frailty (see Fig. 1.4), heart failure is grouped with organ failure [66]. However, only a minority of patients follow the supposedly 'typical' heart failure trajectory [61], and many older patients with heart failure follow a trajectory that more closely mirrors the 'frailty' pattern [67, 68]. Patients with ischemic heart disease are often placed in an 'other' or 'unclassified' group where patients showed a pattern of modest and gradual decline in independence during the final year of life.

Prognostication

Selected tools for prognosticating heart failure patients with end-stage disease are shown in Table 1.1. In the general practice setting, predictors of 1-year mortality include male sex, NYHA class III or IV, age ≥85 years, and cancer [75]. Interestingly, using these broad criteria as a guide, general practitioners were able to identify those who died within 12 months with 79 % sensitivity but only 61 % specificity.

Table 1.1 Prognostic tools to identify heart failure patients approaching end of life

Heart failure		Reported measures of validity
EFFECT Heart Failure Mortality Prediction	In-patient demographic and lab data at admission provide risk stratification for death within 30 days and 1 year (available online at http://www.ccort.ca/Research/CHFRiskModel.aspx) [69]	C-statistic 0.82 for in-hospital death, 0.80 for 30-day mortality and 0.77 for 1-year mortality [70]
The Seattle Heart Failure Model	Demographic, lab, device and medication data provide 1, 2 and 3 year survival estimates (available online at http://depts.washington.edu/shfm/app.php) [71]	C-statistic 0.73
Heart Failure Survival Score	Ambulatory patient risk stratification, including peak vO_2, developed to aid cardiac transplant decision-making [72]	Event-free survival rate at 1 year in the validation sample was 88 ± 4, 60 ± 6, and 35 ± 10 % in the low-, medium-, and high-risk strata, respectively
Gold Standards Framework	4 components: NYHA class III or IV, The surprise question: "Would you be surprised if this patient died in the next 6–12 months?", Repeated heart failure hospitalizations, and difficult physical or psychological symptoms despite optimized tolerated therapy [73]	2 or more of 4 criteria: sensitivity 83%, specificity 22%
RADbound indicators of PAlliative Care Needs (RADPAC)	1. NYHA Class IV 2. >3 hospital admissions per year 3. >3 severe exacerbations of heart failure per year 4. Moderately disabled; dependent. Requires considerable assistance and frequent care (Karnofsky-score ≤50 %) 5. Increasing weight and non-responsive to increased doses of diuretics 6. General deterioration (edema, orthopnea, dyspnea) 7. The patient mentions 'end of life approaching' [74]	N/A

Modified from McKelvie et al. [25]

Challenges of prognostication are also present in specialty clinics. Predictive models based on ambulatory clinic or controlled trial-based cohorts include the Seattle Heart Failure Model (SHFM), the HF Survival Score (HFSS), and the Gold Standards Framework Prognostic indicator (details in Table 1.1). Validation exercises of these models, however, underscore the current state of difficulty with predicting death for those who could be referred to palliative care.

Repeated hospitalization is a significant predictor of adverse prognosis in patients with heart failure [76, 77]. Predictive models based on data at the time of hospitalization, such as the Enhanced Feedback For Effective Cardiac Treatment (EFFECT-HF) risk score [69], which is comprised of simple clinical factors predict short and long-term mortality for heart failure patients with either HFrEF or HFpEF [68, 78]. In a comparison of six different models [69, 71, 79–82] tested in a cohort of patients who ultimately died with heart failure, predicted mortality was highest using the EFFECT-HF model [78]. The subset of covariates, which included serum urea nitrogen, systolic blood pressure, peripheral artery disease, and hyponatremia were especially predictive of mortality [83]. The presence of three or more of these risk factors was associated with 6-month mortality rates as high as 66.7 % [83]. The role of biomarkers such as brain natriuretic peptide, for the identification of patients at end of life has not been delineated.

Objective measures of decreased functional status, or self-reported poor health, may also provide prognostic information in heart failure. Low functional exercise capacity, defined as ≤300 m walked during 6 min, was associated with 1.8-fold increased risk of death, but is a poor standalone indicator of prognosis [84–86]. In addition, low self-reported physical functioning, defined as a score below 25 on the Short Form Health Survey (SF-12), was also shown to be associated with a 1.6-fold increase in risk of death. However, the much more simple measure of poor self-rated general health, corresponded to a 2.7-fold increase in risk of death compared with good to excellent self-reported general health [84]. Functional status information is an important component of the RADbound indicators of Palliative Care Needs and the Gold Standards Framework, which may be useful to identify patients with heart failure approaching end of life (see Table 1.1).

Generic aids to identifying patients approaching end of life include the 'surprise question,' which is a component of the Gold Standards Framework, and the Palliative Performance Scale (see Table 1.2). However, these have not been studied in end-stage heart failure. While older age is associated with a substantially increased risk of death, the frail elderly are at a particularly heightened risk. Few predictive models have included frailty as a parameter, although this metric is rapidly gaining interest. Among frail elderly patients aged 75 years or older with refractory (stage D) heart failure, symptoms of peripheral edema or pain, and need for nitrate therapy were found to predict mortality, whereas the ability to sit in a chair was associated with improved survival [89].

Contrary to common perception, refractory stable angina does not independently predict a high risk of death in the near future. Among those with chronic ischemic heart disease and refractory angina, overall 1-year mortality rate was only 3.9 %, but the

Table 1.2 Generic aids for identifying patients at end of life

Generic	
Surprise question: "Would I be surprised if this patient died in the next 6–12 months?"	Increased odds of dying if answer is "No". Validated in cancer and renal dialysis patients to date [75], and is a component of the Gold Standards Framework
Palliative Performance scale (PPS)	Physical status rating (by clinician) out of 100 % (ability in activity, ambulation, self-care, oral intake and level of consciousness), lower scores associated with poorer prognosis. Developed in cancer, some evidence for non-cancer use (available online at www.victoriahospice.org/sites/default/files/imce/PPS%20ENGLISH.pdf) [87, 88]

Table 1.3 Prognostic tools to identify coronary disease patients approaching end of life

Myocardial infarction		Reported measures of validity
GRACE risk score	Extensively validated risk score among patients with myocardial infarction. Available online at: http://www.outcomes-umassmed.org/GRACE/default.aspx [91, 92]	78 % sensitivity and 89 % specificity for 6-month mortality when combined with Gold Standards framework [93]
Karnofsky Performance status	Functional scale, similar to PPS; one study of use in prognosis with acute MI (http://www.hospicepatients.org/karnofsky.html) [94]	A KPS score <8 (scale from 1 to 10) 3 weeks before the index infarction was associated with a higher incidence of congestive heart failure, in-hospital cardiac arrest, and mortality during hospitalization

presence of diabetes, chronic renal disease, left ventricular dysfunction and heart failure increased the risk of death [90]. While methods have not been developed in those with isolated chronic ischemic heart disease, models for end of life estimation have been proposed for patient subsets with myocardial infarction (see Table 1.3). The overlap in predicted risk factors for death in those with refractory coronary artery disease and end-stage heart failure suggest that a common set of criteria may exist [90].

Impact of Comorbidities on Prognosis

The burden of non-cardiac comorbidities in patients with heart failure has a complementary impact on patients' prognosis. This is increasingly relevant as the population ages, since a larger proportion of patients with heart failure may have accumulated diseases that could have substantial impact on their overall prognosis. From the standpoint of selecting patients for palliation, comorbidities that are minimally-modifiable may be most useful. One of the better studied comorbidities is worsened renal

Table 1.4 Impact of common comorbidities in survival of patients with heart failure

Comorbidity	Prevalence	Impact on prognosis
Renal dysfunction	Any renal impairment: up to 70 % Moderate to severe renal impairment: up to 29 % [95]	15 % increase in risk of death for every 0.5 mg/dl increase in creatinine; 7 % increase for every 10 ml/min decrease in eGFR [95, 96]
Cancer	22 %, with a hazard ratio of 1.68 relative to patients without heart failure [97]	Adjusted hazard ratio of 1.56 for death after adjustment for age, sex, and comorbidities [97]
Anemia	Any anemia 37 % [98]	Adjusted hazard ratio of 1.46 for death with low hemoglobin concentration. Supra-normal hemoglobin values may also be associated with increased risk of death [98–100]

function; however, comorbid cancer and anemia are also common and may not be amenable to curative intervention (see Table 1.4). Other prognostic factors that have been associated with death or persistently unfavorable quality of life include lower systolic blood pressure [22, 23], hyponatremia, tachycardia, and diabetes [101].

Therapeutic Considerations Near End of Life

Increasing diuretic dose may be a useful universal marker of prognosis. In a study of 4,406 elderly patients discharged after heart failure hospitalization [102], the prescription of higher furosemide doses (≥120 mg/day) was more common among patients with higher creatinine levels, diabetes, atrial fibrillation, chronic obstructive pulmonary disease, hypotension, cardiomegaly, hyponatremia, and lower hemoglobin levels. After extensive multivariable adjustment, exposure to higher furosemide dose was found to be predictive of death, hospitalization and renal dysfunction over 5-year follow-up. Intolerance of angiotensin converting enzyme (ACE) inhibitors or beta-adrenoreceptor antagonists is a poor prognostic sign; tolerance of heart failure therapy is one of the predictors of discharge alive from hospice among patients with heart failure [55, 101, 103].

A relevant consideration is the appropriate continuation or termination of medications or device therapy in a patient with advanced cardiovascular disease [104–106]. Patients at the highest risk of adverse outcomes have larger absolute benefit from interventions. However, these patients are often left undertreated resulting in a 'risk-treatment mismatch'. For example, patients with heart failure at greatest risk of death are least likely to receive ACE inhibitors or angiotensin receptor blockers (ARBs), and beta-adrenoreceptor antagonists [104].

The end of life of patients with advanced cardiac disease frequently involves evidence of systemic hypoperfusion, prompting consideration of the use of inotropes. However, the Outcomes of a Prospective Trial of Intravenous Milrinone for

Exacerbations of Chronic Heart Failure (OPTIME-CHF) study demonstrated that inotropes should not be used routinely for prognostic or even symptomatic benefit in unselected patients with heart failure [107]. There may be a select group of 'cold-and-wet' patients, particularly those who are diuretic-resistant, who could benefit from rational inotrope use [108], and for some patients with HFrEF, inotropes may facilitate discharge from hospital to provide the opportunity to die at home in patients with end-stage heart failure [109].

At the end of life, health care professionals may be called upon to make shared decisions regarding resuscitative efforts. In those with end-stage cardiac disease, resuscitation is unsuccessful in 70–98 % of cases [110]; those who do survive a cardiac arrest often experience symptoms of anxiety, depression, post-traumatic stress and difficulties in cognitive function [110, 111]. However, cardiopulmonary resuscitation (CPR) can have favorable outcomes during witnessed shockable rhythms, when performed in the operating room, or during iatrogenic complications of procedures [112].

References

1. World Health Organization. Number of deaths: world by cause. Geneva: World Health Organization; 2011.
2. Singer PA, Bowman KW. Quality end-of-life care: a global perspective. BMC Palliat Care. 2002;1(1):4.
3. World Palliative Care Alliance. Global atlas of palliative care at the end of life. Geneva: World Health Organization; 2014.
4. Murray CJ, Richards MA, Newton JN, Fenton KA, Anderson HR, Atkinson C, Bennett D, Bernabe E, Blencowe H, Bourne R, Braithwaite T, Brayne C, Bruce NG, Brugha TS, Burney P, Dherani M, Dolk H, Edmond K, Ezzati M, Flaxman AD, Fleming TD, Freedman G, Gunnell D, Hay RJ, Hutchings SJ, Ohno SL, Lozano R, Lyons RA, Marcenes W, Naghavi M, Newton CR, Pearce N, Pope D, Rushton L, Salomon JA, Shibuya K, Vos T, Wang H, Williams HC, Woolf AD, Lopez AD, Davis A. UK health performance: findings of the Global Burden of Disease Study 2010. Lancet. 2013;381(9871):997–1020.
5. Murphy SL, Kochanek KD. Deaths: final data for 2010. Hyattsville: National Center for Health Statistics; 2013.
6. Go AS, Mozaffarian D, Roger VL, Benjamin EJ, Berry JD, Blaha MJ, Dai S, Ford ES, Fox CS, Franco S, Fullerton HJ, Gillespie C, Hailpern SM, Heit JA, Howard VJ, Huffman MD, Judd SE, Kissela BM, Kittner SJ, Lackland DT, Lichtman JH, Lisabeth LD, Mackey RH, Magid DJ, Marcus GM, Marelli A, Matchar DB, McGuire DK, Mohler III ER, Moy CS, Mussolino ME, Neumar RW, Nichol G, Pandey DK, Paynter NP, Reeves MJ, Sorlie PD, Stein J, Towfighi A, Turan TN, Virani SS, Wong ND, Woo D, Turner MB. Heart disease and stroke statistics–2014 update: a report from the American Heart Association. Circulation. 2014;129(3):e28–292.
7. British National Health Service. Deaths from cardiovascular diseases: implications for end of life care in England. National End of Life Care Intelligence Network; 2013.
8. Chun S, Tu JV, Wijeysundera HC, Austin PC, Wang X, Levy D, Lee DS. Lifetime analysis of hospitalizations and survival of patients newly admitted with heart failure. Circ Heart Fail. 2012;5(4):414–21.
9. Levy D, Kenchaiah S, Larson MG, Benjamin EJ, Kupka MJ, Ho KK, Murabito JM, Vasan RS. Long-term trends in the incidence of and survival with heart failure. N Engl J Med. 2002;347(18):1397–402.

10. Roger VL, Weston SA, Redfield MM, Hellermann-Homan JP, Killian J, Yawn BP, Jacobsen SJ. Trends in heart failure incidence and survival in a community-based population. JAMA. 2004;292(3):344–50.
11. Owan TE, Hodge DO, Herges RM, Jacobsen SJ, Roger VL, Redfield MM. Trends in prevalence and outcome of heart failure with preserved ejection fraction. N Engl J Med. 2006;355(3):251–9.
12. Bhatia RS, Tu JV, Lee DS, Austin PC, Fang J, Haouzi A, Gong Y, Liu PP. Outcome of heart failure with preserved ejection fraction in a population-based study. N Engl J Med. 2006;355(3):260–9.
13. Fonarow GC, Stough WG, Abraham WT, Albert NM, Gheorghiade M, Greenberg BH, O'Connor CM, Sun JL, Yancy CW, Young JB. Characteristics, treatments, and outcomes of patients with preserved systolic function hospitalized for heart failure: a report from the OPTIMIZE-HF Registry. J Am Coll Cardiol. 2007;50(8):768–77.
14. Lee DS, Gona P, Albano I, Larson MG, Benjamin EJ, Levy D, Kannel WB, Vasan RS. A systematic assessment of causes of death after heart failure onset in the community: impact of age at death, time period, and left ventricular systolic dysfunction. Circ Heart Fail. 2011;4(1):36–43.
15. Henkel DM, Redfield MM, Weston SA, Gerber Y, Roger VL. Death in heart failure: a community perspective. Circ Heart Fail. 2008;1(2):91–7.
16. Unroe KT, Greiner MA, Hernandez AF, Whellan DJ, Kaul P, Schulman KA, Peterson ED, Curtis LH. Resource use in the last 6 months of life among medicare beneficiaries with heart failure, 2000–2007. Arch Intern Med. 2011;171(3):196–203.
17. Kaul P, McAlister FA, Ezekowitz JA, Bakal JA, Curtis LH, Quan H, Knudtson ML, Armstrong PW. Resource use in the last 6 months of life among patients with heart failure in Canada. Arch Intern Med. 2011;171(3):211–7.
18. Hall MJ, Levant S, DeFrances CJ. Hospitalization for congestive heart failure: United States, 2000–2010. NCHS data brief no 108. Hyattsville: National Center for Health Statistics; 2012.
19. Riegel B, Moser DK, Powell M, Rector TS, Havranek EP. Nonpharmacologic care by heart failure experts. J Card Fail. 2006;12(2):149–53.
20. Kirkpatrick JN, Kim AY. Ethical issues in heart failure: overview of an emerging need. Perspect Biol Med. 2006;49(1):1–9.
21. Setoguchi S, Glynn RJ, Stedman M, Flavell CM, Levin R, Stevenson LW. Hospice, opiates, and acute care service use among the elderly before death from heart failure or cancer. Am Heart J. 2010;160(1):139–44.
22. Hauptman PJ, Goodlin SJ, Lopatin M, Costanzo MR, Fonarow GC, Yancy CW. Characteristics of patients hospitalized with acute decompensated heart failure who are referred for hospice care. Arch Intern Med. 2007;167(18):1990–7.
23. Whellan DJ, Cox M, Hernandez AF, Heidenreich PA, Curtis LH, Peterson ED, Fonarow GC. Utilization of hospice and predicted mortality risk among older patients hospitalized with heart failure: findings from GWTG-HF. J Card Fail. 2012;18(6):471–7.
24. Hunt SA, Abraham WT, Chin MH, Feldman AM, Francis GS, Ganiats TG, Jessup M, Konstam MA, Mancini DM, Michl K, Oates JA, Rahko PS, Silver MA, Stevenson LW, Yancy CW. 2009 focused update incorporated into the ACC/AHA 2005 Guidelines for the Diagnosis and Management of Heart Failure in Adults: a report of the American College of Cardiology Foundation/American Heart Association Task Force on Practice Guidelines: developed in collaboration with the International Society for Heart and Lung Transplantation. J Am Coll Cardiol. 2009;53(15):e1–90.
25. McKelvie RS, Moe GW, Cheung A, Costigan J, Ducharme A, Estrella-Holder E, Ezekowitz JA, Floras J, Giannetti N, Grzeslo A, Harkness K, Heckman GA, Howlett JG, Kouz S, Leblanc K, Mann E, O'Meara E, Rajda M, Rao V, Simon J, Swiggum E, Zieroth S, Arnold JM, Ashton T, D'Astous M, Dorian P, Haddad H, Isaac DL, Leblanc MH, Liu P, Sussex B, Ross HJ. The 2011 Canadian Cardiovascular Society heart failure management guidelines update: focus on sleep apnea, renal dysfunction, mechanical circulatory support, and palliative care. Can J Cardiol. 2011;27(3):319–38.

26. Jaarsma T, Beattie JM, Ryder M, Rutten FH, McDonagh T, Mohacsi P, Murray SA, Grodzicki T, Bergh I, Metra M, Ekman I, Angermann C, Leventhal M, Pitsis A, Anker SD, Gavazzi A, Ponikowski P, Dickstein K, Delacretaz E, Blue L, Strasser F, McMurray J. Palliative care in heart failure: a position statement from the palliative care workshop of the Heart Failure Association of the European Society of Cardiology. Eur J Heart Fail. 2009;11(5):433–43.
27. Givens JL, Tjia J, Zhou C, Emanuel E, Ash AS. Racial and ethnic differences in hospice use among patients with heart failure. Arch Intern Med. 2010;170(5):427–32.
28. Green E, Gardiner C, Gott M, Ingleton C. Exploring the extent of communication surrounding transitions to palliative care in heart failure: the perspectives of health care professionals. J Palliat Care. 2011;27(2):107–16.
29. Howlett J, Morin L, Fortin M, Heckman G, Strachan PH, Suskin N, Shamian J, Lewanczuk R, Aurthur HM. End-of-life planning in heart failure: it should be the end of the beginning. Can J Cardiol. 2010;26(3):135–41.
30. Momen NC, Barclay SI. Addressing 'the elephant on the table': barriers to end of life care conversations in heart failure – a literature review and narrative synthesis. Curr Opin Support Palliat Care. 2011;5(4):312–6.
31. Murray SA, Boyd K, Kendall M, Worth A, Benton TF, Clausen H. Dying of lung cancer or cardiac failure: prospective qualitative interview study of patients and their careers in the community. BMJ. 2002;325(7370):929.
32. Bekelman DB, Rumsfeld JS, Havranek EP, Yamashita TE, Hutt E, Gottlieb SH, Dy SM, Kutner JS. Symptom burden, depression, and spiritual well-being: a comparison of heart failure and advanced cancer patients. J Gen Intern Med. 2009;24(5):592–8.
33. Janssen DJ, Spruit MA, Wouters EF, Schols JM. Daily symptom burden in end-stage chronic organ failure: a systematic review. Palliat Med. 2008;22(8):938–48.
34. Ekman I, Cleland JG, Swedberg K, Charlesworth A, Metra M, Poole-Wilson PA. Symptoms in patients with heart failure are prognostic predictors: insights from COMET. J Card Fail. 2005;11(4):288–92.
35. Sherwood A, Blumenthal JA, Trivedi R, Johnson KS, O'Connor CM, Adams Jr KF, Dupree CS, Waugh RA, Bensimhon DR, Gaulden L, Christenson RH, Koch GG, Hinderliter AL. Relationship of depression to death or hospitalization in patients with heart failure. Arch Intern Med. 2007;167(4):367–73.
36. Steptoe A, Molloy GJ, Messerli-Burgy N, Wikman A, Randall G, Perkins-Porras L, Kaski JC. Fear of dying and inflammation following acute coronary syndrome. Eur Heart J. 2011;32(19):2405–11.
37. Ekman I, Ehrenberg A. Fatigued elderly patients with chronic heart failure: do patient reports and nurse recordings correspond? Int J Nurs Terminol Classif. 2002;13(4):127–36.
38. Gadsboll N, Hoilund-Carlsen PF, Nielsen GG, Berning J, Brunn NE, Stage P, Hein E, Marving J, Longborg-Jensen H, Jensen BH. Symptoms and signs of heart failure in patients with myocardial infarction: reproducibility and relationship to chest X-ray, radionuclide ventriculography and right heart catheterization. Eur Heart J. 1989;10(11):1017–28.
39. Ekman I, Cleland JG, Andersson B, Swedberg K. Exploring symptoms in chronic heart failure. Eur J Heart Fail. 2005;7(5):699–703.
40. Zambroski CH, Moser DK, Bhat G, Ziegler C. Impact of symptom prevalence and symptom burden on quality of life in patients with heart failure. Eur J Cardiovasc Nurs. 2005;4(3):198–206.
41. Opasich C, Gualco A, De FS, Barbieri M, Cioffi G, Giardini A, Majani G. Physical and emotional symptom burden of patients with end-stage heart failure: what to measure, how and why. J Cardiovasc Med (Hagerstown). 2008;9(11):1104–8.
42. Walke LM, Byers AL, McCorkle R, Fried TR. Symptom assessment in community-dwelling older adults with advanced chronic disease. J Pain Symptom Manage. 2006;31(1):31–7.
43. Steinhauser KE, Clipp EC, Bosworth HB, McNeilly M, Christakis NA, Voils CI, Tulsky JA. Measuring quality of life at the end of life: validation of the QUAL-E. Palliat Support Care. 2004;2(1):3–14.
44. Cohen SR, Mount BM, Strobel MG, Bui F. The McGill Quality of Life Questionnaire: a measure of quality of life appropriate for people with advanced disease. A preliminary study of validity and acceptability. Palliat Med. 1995;9(3):207–19.

45. Ezekowitz JA, Thai V, Hodnefield TS, Sanderson L, Cujec B. The correlation of standard heart failure assessment and palliative care questionnaires in a multidisciplinary heart failure clinic. J Pain Symptom Manage. 2011;42(3):379–87.
46. Schwarz ER, Baraghoush A, Morrissey RP, Shah AB, Shinde AM, Phan A, Bharadwaj P. Pilot study of palliative care consultation in patients with advanced heart failure referred for cardiac transplantation. J Palliat Med. 2012;15(1):12–5.
47. Paes P. A pilot study to assess the effectiveness of a palliative care clinic in improving the quality of life for patients with severe heart failure. Palliat Med. 2005;19(6):505–6.
48. Connor SR, Pyenson B, Fitch K, Spence C, Iwasaki K. Comparing hospice and nonhospice patient survival among patients who die within a three-year window. J Pain Symptom Manage. 2007;33(3):238–46.
49. Johnson MJ, Houghton T. Palliative care for patients with heart failure: description of a service. Palliat Med. 2006;20(3):211–4.
50. Selman L, Harding R, Beynon T, Hodson F, Hazeldine C, Coady E, Gibbs L, Higginson IJ. Modelling services to meet the palliative care needs of chronic heart failure patients and their families: current practice in the UK. Palliat Med. 2007;21(5):385–90.
51. Rutten FH, Heddema WS, Daggelders GJ, Hoes AW. Primary care patients with heart failure in the last year of their life. Fam Pract. 2012;29(1):36–42.
52. Blecker S, Anderson GF, Herbert R, Wang NY, Brancati FL. Hospice care and resource utilization in Medicare beneficiaries with heart failure. Med Care. 2011;49(11):985–91.
53. Emanuel EJ. Cost savings at the end of life. What do the data show? JAMA. 1996;275(24): 1907–14.
54. Taylor Jr DH, Ostermann J, Van Houtven CH, Tulsky JA, Steinhauser K. What length of hospice use maximizes reduction in medical expenditures near death in the US Medicare program? Soc Sci Med. 2007;65(7):1466–78.
55. Bain KT, Maxwell TL, Strassels SA, Whellan DJ. Hospice use among patients with heart failure. Am Heart J. 2009;158(1):118–25.
56. Bekelman DB, Nowels CT, Allen LA, Shakar S, Kutner JS, Matlock DD. Outpatient palliative care for chronic heart failure: a case series. J Palliat Med. 2011;14(7):815–21.
57. Adler ED, Goldfinger JZ, Kalman J, Park ME, Meier DE. Palliative care in the treatment of advanced heart failure. Circulation. 2009;120(25):2597–606.
58. Goodlin SJ. End-of-life care in heart failure. Curr Cardiol Rep. 2009;11(3):184–91.
59. O'Leary N. The comparative palliative care needs of those with heart failure and cancer patients. Curr Opin Support Palliat Care. 2009;3(4):241–6.
60. Gibbs LM, Khatri AK, Gibbs JS. Survey of specialist palliative care and heart failure: September 2004. Palliat Med. 2006;20(6):603–9.
61. Gott M, Barnes S, Parker C, Payne S, Seamark D, Gariballa S, Small N. Dying trajectories in heart failure. Palliat Med. 2007;21(2):95–9.
62. Johnson MJ, Gadoud A. Palliative care for people with chronic heart failure: when is it time? J Palliat Care. 2011;27(1):37–42.
63. Allen LA, Yager JE, Funk MJ, Levy WC, Tulsky JA, Bowers MT, Dodson GC, O'Connor CM, Felker GM. Discordance between patient-predicted and model-predicted life expectancy among ambulatory patients with heart failure. JAMA. 2008;299(21):2533–42.
64. Murray SA, Kendall M, Boyd K, Sheikh A. Illness trajectories and palliative care. BMJ. 2005;330(7498):1007–11.
65. Lunney JR, Lynn J, Hogan C. Profiles of older medicare decedents. J Am Geriatr Soc. 2002;50(6):1108–12.
66. Lunney JR, Lynn J, Foley DJ, Lipson S, Guralnik JM. Patterns of functional decline at the end of life. JAMA. 2003;289(18):2387–92.
67. Chen JH, Chan DC, Kiely DK, Morris JN, Mitchell SL. Terminal trajectories of functional decline in the long-term care setting. J Gerontol A Biol Sci Med Sci. 2007;62(5):531–6.
68. Ko DT, Alter DA, Austin PC, You JJ, Lee DS, Qiu F, Stukel TA, Tu JV. Life expectancy after an index hospitalization for patients with heart failure: a population-based study. Am Heart J. 2008;155(2):324–31.

69. Lee DS, Austin PC, Rouleau JL, Liu PP, Naimark D, Tu JV. Predicting mortality among patients hospitalized for heart failure: derivation and validation of a clinical model. JAMA. 2003;290(19):2581–7.
70. Lee DS, Ezekowitz JA. Risk stratification in acute heart failure. Can J Cardiol. 2014;30(3):312–9.
71. Levy WC, Mozaffarian D, Linker DT, Sutradhar SC, Anker SD, Cropp AB, Anand I, Maggioni A, Burton P, Sullivan MD, Pitt B, Poole-Wilson PA, Mann DL, Packer M. The Seattle Heart Failure Model: prediction of survival in heart failure. Circulation. 2006;113(11):1424–33.
72. Aaronson KD, Schwartz JS, Chen TM, Wong KL, Goin JE, Mancini DM. Development and prospective validation of a clinical index to predict survival in ambulatory patients referred for cardiac transplant evaluation. Circulation. 1997;95(12):2660–7.
73. Haga K, Murray S, Reid J, Ness A, O'Donnell M, Yellowlees D, Denvir MA. Identifying community based chronic heart failure patients in the last year of life: a comparison of the Gold Standards Framework Prognostic Indicator Guide and the Seattle Heart Failure Model. Heart. 2012;98(7):579–83.
74. Thoonsen B, Engels Y, van Rijswijk E, Verhagen S, van Weel C, Groot M, Vissers K. Early identification of palliative care patients in general practice: development of RADboud indicators for PAlliative Care Needs (RADPAC). Br J Gen Pract. 2012;62(602):e625–31.
75. Barnes S, Gott M, Payne S, Parker C, Seamark D, Gariballa S, Small N. Predicting mortality among a general practice-based sample of older people with heart failure. Chronic Illn. 2008;4(1):5–12.
76. Lee DS, Austin PC, Stukel TA, Alter DA, Chong A, Parker JD, Tu JV. "Dose-dependent" impact of recurrent cardiac events on mortality in patients with heart failure. Am J Med. 2009;122(2):162–9.
77. Setoguchi S, Stevenson LW, Schneeweiss S. Repeated hospitalizations predict mortality in the community population with heart failure. Am Heart J. 2007;154(2):260–6.
78. Nutter AL, Tanawuttiwat T, Silver MA. Evaluation of 6 prognostic models used to calculate mortality rates in elderly heart failure patients with a fatal heart failure admission. Congest Heart Fail. 2010;16(5):196–201.
79. Aaronson KD, Schwartz JS, Chen T, Wong K, Goin JE, Mancini DM. Development and prospective validation of a clinical index to predict survival in ambulatory patients referred for cardiac transplant evaluation. Circulation. 1997;95(12):2660–7.
80. Fonarow GC, Adams Jr KF, Abraham WT, Yancy CW, Boscardin WJ. Risk stratification for in-hospital mortality in acutely decompensated heart failure: classification and regression tree analysis. JAMA. 2005;293(5):572–80.
81. Heywood JT, Elatre W, Pai RG, Fabbri S, Huiskes B. Simple clinical criteria to determine the prognosis of heart failure. J Cardiovasc Pharmacol Ther. 2005;10(3):173–80.
82. Pocock SJ, Wang D, Pfeffer MA, Yusuf S, McMurray JJ, Swedberg KB, Ostergren J, Michelson EL, Pieper KS, Granger CB. Predictors of mortality and morbidity in patients with chronic heart failure. Eur Heart J. 2006;27(1):65–75.
83. Huynh BC, Rovner A, Rich MW. Identification of older patients with heart failure who may be candidates for hospice care: development of a simple four-item risk score. J Am Geriatr Soc. 2008;56(6):1111–5.
84. Chamberlain AM, McNallan SM, Dunlay SM, Spertus JA, Redfield MM, Moser DK, Kane RL, Weston SA, Roger VL. Physical health status measures predict all-cause mortality in patients with heart failure. Circ Heart Fail. 2013;6(4):669–75.
85. Guazzi M, Dickstein K, Vicenzi M, Arena R. Six-minute walk test and cardiopulmonary exercise testing in patients with chronic heart failure: a comparative analysis on clinical and prognostic insights. Circ Heart Fail. 2009;2(6):549–55.
86. Opasich C, Pinna GD, Mazza A, Febo O, Riccardi R, Riccardi PG, Capomolla S, Forni G, Cobelli F, Tavazzi L. Six-minute walking performance in patients with moderate-to-severe heart failure; is it a useful indicator in clinical practice? Eur Heart J. 2001;22(6):488–96.
87. Head B, Ritchie CS, Smoot TM. Prognostication in hospice care: can the palliative performance scale help? J Palliat Med. 2005;8(3):492–502.

88. Harrold J, Rickerson E, Carroll JT, McGrath J, Morales K, Kapo J, Casarett D. Is the palliative performance scale a useful predictor of mortality in a heterogeneous hospice population? J Palliat Med. 2005;8(3):503–9.
89. Martin-Pfitzenmeyer I, Gauthier S, Bailly M, Loi N, Popitean L, d'Athis P, Bouvier AM, Pfitzenmeyer P. Prognostic factors in stage D heart failure in the very elderly. Gerontology. 2009;55(6):719–26.
90. Henry TD, Satran D, Hodges JS, Johnson RK, Poulose AK, Campbell AR, Garberich RF, Bart BA, Olson RE, Boisjolie CR, Harvey KL, Arndt TL, Traverse JH. Long-term survival in patients with refractory angina. Eur Heart J. 2013;34:2683–8.
91. Granger CB, Goldberg RJ, Dabbous O, Pieper KS, Eagle KA, Cannon CP, Van De Werf F, Avezum A, Goodman SG, Flather MD, Fox KA. Predictors of hospital mortality in the global registry of acute coronary events. Arch Intern Med. 2003;163(19):2345–53.
92. Eagle KA, Lim MJ, Dabbous OH, Pieper KS, Goldberg RJ, Van De Werf F, Goodman SG, Granger CB, Steg PG, Gore JM, Budaj A, Avezum A, Flather MD, Fox KA. A validated prediction model for all forms of acute coronary syndrome: estimating the risk of 6-month postdischarge death in an international registry. JAMA. 2004;291(22):2727–33.
93. Fenning S, Woolcock R, Haga K, Iqbal J, Fox KA, Murray SA, Denvir MA. Identifying acute coronary syndrome patients approaching end-of-life. PLoS One. 2012;7(4):e35536.
94. Brezinski D, Stone PH, Muller JE, Tofler GH, Davis V, Parker C, Hartley LH, Braunwald E. Prognostic significance of the Karnofsky Performance Status score in patients with acute myocardial infarction: comparison with the left ventricular ejection fraction and the exercise treadmill test performance. The MILIS Study Group. Am Heart J. 1991;121(5):1374–81.
95. Smith GL, Lichtman JH, Bracken MB, Shlipak MG, Phillips CO, DiCapua P, Krumholz HM. Renal impairment and outcomes in heart failure: systematic review and meta-analysis. J Am Coll Cardiol. 2006;47(10):1987–96.
96. Hillege HL, Nitsch D, Pfeffer MA, Swedberg K, McMurray JJ, Yusuf S, Granger CB, Michelson EL, Ostergren J, Cornel JH, de Zeeuw D, Pocock S, van Veldhuisen DJ. Renal function as a predictor of outcome in a broad spectrum of patients with heart failure. Circulation. 2006;113(5):671–8.
97. Hasin T, Gerber Y, McNallan SM, Weston SA, Kushwaha SS, Nelson TJ, Cerhan JR, Roger VL. Patients with heart failure have an increased risk of incident cancer. J Am Coll Cardiol. 2013;62(10):881–6.
98. Groenveld HF, Januzzi JL, Damman K, van Wijngaarden J, Hillege HL, van Veldhuisen DJ, van der Meer P. Anemia and mortality in heart failure patients a systematic review and meta-analysis. J Am Coll Cardiol. 2008;52(10):818–27.
99. Komajda M, Anker SD, Charlesworth A, Okonko D, Metra M, Di LA, Remme W, Moullet C, Swedberg K, Cleland JG, Poole-Wilson PA. The impact of new onset anaemia on morbidity and mortality in chronic heart failure: results from COMET. Eur Heart J. 2006;27(12):1440–6.
100. Anand IS, Kuskowski MA, Rector TS, Florea VG, Glazer RD, Hester A, Chiang YT, Aknay N, Maggioni AP, Opasich C, Latini R, Cohn JN. Anemia and change in hemoglobin over time related to mortality and morbidity in patients with chronic heart failure: results from Val-HeFT. Circulation. 2005;112(8):1121–7.
101. Allen LA, Gheorghiade M, Reid KJ, Dunlay SM, Chan PS, Hauptman PJ, Zannad F, Konstam MA, Spertus JA. Identifying patients hospitalized with heart failure at risk for unfavorable future quality of life. Circ Cardiovasc Qual Outcomes. 2011;4(4):389–98.
102. Abdel-Qadir HM, Tu JV, Yun L, Austin PC, Newton GE, Lee DS. Diuretic dose and long-term outcomes in elderly patients with heart failure after hospitalization. Am Heart J. 2010;160(2):264–71.
103. Berger AK, Duval S, Manske C, Vazquez G, Barber C, Miller L, Luepker RV. Angiotensin-converting enzyme inhibitors and angiotensin receptor blockers in patients with congestive heart failure and chronic kidney disease. Am Heart J. 2007;153(6):1064–73.
104. Lee DS, Tu JV, Juurlink DN, Alter DA, Ko DT, Austin PC, Chong A, Stukel TA, Levy D, Laupacis A. Risk-treatment mismatch in the pharmacotherapy of heart failure. JAMA. 2005;294(10):1240–7.

105. Stevenson JP, Currow DC, Abernethy AP. The broader implications of managing statins at the end of life. J Palliat Care. 2007;23(3):188–9.
106. Tanvetyanon T, Leighton JC. Life-sustaining treatments in patients who died of chronic congestive heart failure compared with metastatic cancer. Crit Care Med. 2003;31(1):60–4.
107. Cuffe MS, Califf RM, Adams Jr KF, Benza R, Bourge R, Colucci WS, Massie BM, O'Connor CM, Pina I, Quigg R, Silver MA, Gheorghiade M. Short-term intravenous milrinone for acute exacerbation of chronic heart failure: a randomized controlled trial. JAMA. 2002;287(12):1541–7.
108. Felker GM, O'Connor CM. Inotropic therapy for heart failure: an evidence-based approach. Am Heart J. 2001;142(3):393–401.
109. Taitel M, Meaux N, Pegus C, Valerian C, Kirkham H. Place of death among patients with terminal heart failure in a continuous inotropic infusion program. Am J Hosp Palliat Care. 2012;29(4):249–53.
110. Lippert FK, Raffay V, Georgiou M, Steen PA, Bossaert L. European Resuscitation Council Guidelines for Resuscitation 2010 Section 10. The ethics of resuscitation and end-of-life decisions. Resuscitation. 2010;81(10):1445–51.
111. Pearlman RA, Cain KC, Patrick DL, Appelbaum-Maizel M, Starks HE, Jecker NS, Uhlmann RF. Insights pertaining to patient assessments of states worse than death. J Clin Ethics. 1993;4(1):33–41.
112. Choudhry NK, Choudhry S, Singer PA. CPR for patients labeled DNR: the role of the limited aggressive therapy order. Ann Intern Med. 2003;138(1):65–8.

Chapter 2
Decision Making About End of Life Care: Advance Directives, Durable Power of Attorney for Healthcare, and Talking with Patients with Heart Disease About Dying

Craig Tanner

Abstract Advancing technology has complicated decision making about end of life care. Advance directives arose as a legal response to concerns about patients receiving unwanted interventions near the end of life. Healthcare advance directives and durable power of attorney for healthcare held the promise of controlling medical care at the end of life. This promise has not been realized. All clinicians providing care to patients with heart failure or other life-limiting illnesses should develop a set of communication skills to discuss goals of care, resuscitation preferences and wishes for end of life care. Patient centered communication skills and empathetic response to emotion are the foundational tools for addressing these issues. Newer models of advance care planning are more based on communication between providers, patients and families than on legal documents The POLST paradigm represents an important innovation in allowing patients to document their preferences for care in a portable document which is a legal physician order.

Keywords Advance directives • Advance care planning • End of life • Palliative care • POLST • Resuscitation • Communication skills • Empathy • Living will • Power of attorney • Durable power of attorney for healthcare

---

C. Tanner, MD
General Medicine and Geriatrics, Portland VA Medical Center,
3710 SW US Veterans Hospital Rd (P3MED), Portland, OR 97239, USA

Medicine Department, Oregon Health & Science University, Portland, OR, USA
e-mail: tannercr@ohsu.edu, craig.tanner2@va.gov

S.J. Goodlin, M.W. Rich (eds.), *End-of-Life Care in Cardiovascular Disease*,
DOI 10.1007/978-1-4471-6521-7_2

Key Points

- Advance Directives (AD) were initially developed as legal tools allowing patients to refuse life-sustaining treatment
- Traditional AD have failed to provide patients with ability to control their care when they cannot speak for themselves
- Advance Care Planning (ACP) refers to a broader set of interventions including communication with patients/families as well as documents specifying preferences for care.
- Communication in ACP should be patient/family-centered and focus on patient/family values, honest discussion of illness, addressing strong emotions and shared decision-making
- Newer models of ACP focus more on communication between patients, their families and care providers
- POLST is a promising tool allowing patients with advanced illness to document preferences for aggressiveness of care in a legal physician order

Introduction

As technology and care for heart disease have advanced, planning for and making decisions about end-of-life care has become increasingly complicated. While the foundational principles of medical ethics (autonomy, beneficence, non-maleficence and justice) still guide the provision of medical care, their application has been made more difficult by our increasing ability to keep individuals alive in health states that would not have been possible even in previous generations. The initial response to advancing technology came from the legal world in an attempt to prevent patients from undergoing unwanted interventions. This legal approach now represents only one facet of a larger group of tools and techniques for helping patients and their families plan for the end of life. This chapter will describe the development of advance directives, their pitfalls and promise in the context emerging models for advance care planning and address skills clinicians can use in communicating with and caring for their patients dying with heart disease.

Brief History of Advance Directives and Durable Power of Attorney for Healthcare

The first healthcare advance directive (AD) was created in 1967 by Luis Kutner, a Chicago attorney representing the Euthanasia Society of America. Kutner referred to the document as a “living will”, a “declaration determining the termination of life” and a “testament permitting death” among other names, and based the concept on the fact that “the law provides that a patient may not be subjected to treatment

Table 2.1 Legal history of advance directives and durable power of attorney for healthcare in the United States

Year	Event
1967	First advance directive created by Luis Kutner
1976	First state law (California) legally sanctioning living wills
1976	New Jersey Supreme Court establishes right of family member to act as healthcare proxy (Karen Ann Quinlan Case)
1983	First state law (California) establishing durable power of attorney for healthcare
1991	U.S. Congress passes the Patient Self-Determination Act requiring healthcare organizations to provide information to adult patients about advance healthcare directives upon admission
1997	All states and the District of Columbia have laws establishing durable power of attorney for healthcare

without his consent" [1]. Table 2.1 contains a timeline of important legal events in the history AD and durable power of attorney for healthcare. A detailed history of AD, durable power of attorney for healthcare and advance care planning has previously been published [2].

The Failure of Traditional Advance Directives

The promise of the health care AD was that individuals could retain some control over their future healthcare and the medical aspects of the end of their lives. Unfortunately this promise has largely gone unfulfilled. The factors contributing to this failure have been described by many authors and include the following [3–5]:

- Relatively few patients complete an AD
- The instructions provided in AD documents are often vague and difficult to relate to actual scenarios.
- Even when patients have completed an AD their healthcare providers are often unaware of them.
- Even when present, an AD does not appear to affect actual care at the end of life
- Surrogates named in an AD seldom understand the patient's wishes
- Autonomy may be over-valued as a guiding principle, particularly in western societies.
- Planning for future care is often seen as a one-time event when an AD is completed.

Several of the factors above have been attributed to these documents' origins as legal tools used to assert a right to refuse life-prolonging treatment in certain circumstances. In contrast to the legal approach to AD, a communications approach' to advance directives addresses the broader concept of advance care planning (ACP) [2]. ACP is a concept which encompasses not only completion of certain legal documents specifying preferences and/or naming a surrogate decision-maker, but also ongoing communication between healthcare providers, patients and patients' families. ACP involves addressing patients' and families' concerns about their future health as well as financial, spiritual and other issues related to care at the end of life. Furthermore, ACP involves ongoing communication between providers, patients and their families as their disease progresses. This is particularly important, as preferences for care have been demonstrated to change over time [6].

Talking with Patients About Dying with Heart Disease

In their landmark 2001 report on healthcare quality in the twenty-first century the Institute of Medicine (IOM) stated that quality healthcare should be [7]:

> Patient-centered—providing care that is respectful of and responsive to individual patient preferences, needs and values and ensuring that patient values guide all decisions.

Multiple studies have demonstrated that patients with chronic illnesses such as heart failure frequently want to discuss prognosis and planning for future care with their healthcare providers. Furthermore there is evidence that patients with the opportunity to have such discussions with their physician are more likely to receive end-of-life care consistent with their preferences. Unfortunately there is also evidence that opportunities to have such conversations are frequently missed. The reasons for this disconnect are complex and likely include provider, patient, family and system barriers such as [8]:

Patient barriers: Cognitive impairment, low health literacy, fear, uncontrolled symptoms, depression, language/cultural barriers.

Family barriers: Family conflict, no family available, cultural/language barriers

System barriers: Providers may not be reimbursed for a dedicated visit to discuss goals of care. Providers may not be trained to competently conduct such discussions.

Provider barriers: Physicians and other healthcare providers often lack the knowledge, skills and confidence to discuss goals of care with their patients with chronic heart failure and other chronic illnesses. Communication skills are recognized as core skills in providing patient-centered care, however many clinicians feel unprepared in this area. Providers also describe time constraints as a significant barrier to discussion of goals of care [8]. Cultural and ethnic differences can also pose barriers to effective discussion of goals of care and effective end of life care [9].

Separate studies have reported that heart failure populations have a survival rate 5-years after diagnosis or first hospitalization in the range of 25–35 % putting the disease on par with stage IIIa ovarian cancer and with a worse prognosis than stage IIIb colorectal cancer [10, 11]. Unfortunately clinicians frequently miss opportunities to discuss prognosis, goals of care and advance care planning [12].

All practitioners should develop a repertoire of communication skills for navigating the discussion of diagnosis, prognosis, goals of care, treatment preferences (including resuscitation preferences) as well as values, hopes and expectations for the future.

Table 2.2 Communication skills

Tool/task	Possible questions/statements
Ask	"What do you understand about your illness?"
	"What have you been told about your illness?"
Tell	"I'm sorry to tell you that your heart has gotten worse"
	"The test results were not what we were hoping for"
Ask	"Sometimes I don't explain things clearly, can you tell me what you heard in this conversation?"
	"What questions do you have?"
N-U-R-S-E—responding to emotion	Suggested language
N-Name the emotion	"Some people would be angry at this news"
U-Understand	"This must be very difficult..."
R-Respect (praise)	"I'm very impressed by how you've coped with..."
S-Support	"Our team with be with you throughout this"
E-Explore	"Tell me more..."

Communication Tools for Talking with Patients About Living and Dying with Heart Disease

Two fundamental communication tools advocated for advance care planning discussions are outlined in Table 2.2 [13]. These tools will be used in many of the examples to follow and will be highlighted when used. The Ask-Tell-Ask model has been described previously and provides a fundamental framework for discussing difficult issues with patients and their families. The N-U-R-S-E mnemonic is one method of describing empathetic responses to emotions that arise during discussions. This mnemonic itself is less important than developing a repertoire of skills and statements to use in responding to emotion during difficult conversations.

Common Communication Challenges in Patients with Heart Failure

Discussing Serious News/Discussing Prognosis and Uncertainty

Heart failure is a chronic illness with a 5-year survival rate on par with many advanced malignancies. Discussing a new diagnosis or the expected future course of heart failure allows patients and their families to plan for their futures as well as to receive care that is consistent with their goals and values. Even though the trajectory of heart failure can be widely variable and prognosticating is difficult and inaccurate, all patients should have the opportunity to discuss their diagnosis and prognosis to

Table 2.3 Discussing diagnosis and prognosis

Step	Example of suggested language
Prepare setting	"Who else would you like to have here while we discuss this?"
Assess patient/family understanding (*Ask*)	"What do you know (or what have you been told) about your illness?"
Describe diagnosis using easy to understand language (*Tell*)	"You have heart failure. Your heart is having trouble pumping blood throughout your body. "
	"Your kidneys aren't getting enough blood"
Give clear, honest prognostic information, remaining hopeful	"Heart failure does shorten people's lives. The good news is that there are treatments that can help you feel better and live longer"
Acknowledge Uncertainty	"I can't say what will happen to you, I can give you some averages from what we see in patients with heart failure"
	"I wish I could predict more exactly"
Respond to Emotion with Empathy	"Most people find this conversation upsetting"
	"These are difficult topics"
	"I'm impressed by how well you've coped with the changes in your life"
Assess Patient/Family Understanding (*Ask)*	"Tell me what you understand about what I said, so I can be sure I was clear"
	"What questions do you have about what we have discussed?"
Make plans for follow-up and commit to ongoing care	"We will continue these medications for now."
	"I will bring these issues up periodically. You should also do so if you wish to discuss them."
	"I will keep working with you to help you live as long and well as possible"

the extent they desire [14]. Table 2.3 outlines a suggested approach to discussion of diagnosis and prognosis with suggested language and explanation. Some patients will prefer not to receive prognostic information, and this preference should be honored if at all possible. If such information is needed to ensure informed consent a patient can delegate decision-making to a family member or friend. Practitioners should develop a repertoire of communication skills for discussing prognosis, uncertainty and serious news or changes in condition, and such issues should be revisited regularly rather addressed as a single discussion.

Discussing Resuscitation Preferences

Discussion of patient preferences in the event of cardiac arrest should begin early in the course of heart failure and continue on an ongoing basis. Table 2.4 presents the framework, with suggested language, for discussion of resuscitation preferences. Physicians should not hesitate to make a recommendation about resuscitation status and other treatment [15]. Such recommendations should always be based on other

Table 2.4 Discussing resuscitation preferences

Step	Suggested language
Set Context-Normalize Topic	"I talk with all of my patients with heart failure about this topic"
Use straightforward language, use "when", not "if"	"When your heart stops do you want us to try to revive you or let you die naturally"?
Explore Patient Reasoning	"Have you had experiences with resuscitation for yourself or a friend or family member?"
	"How do you and your family make important decisions?"
Identify Undesired States	"Are there any health states in which you would not want life-prolonging treatments?"
	"Unable to communicate with your family/friends?"
	"Unable to care for yourself?"
	"Unable to live independently?"
Discuss preferences for location or level of care or living situation	"Would you want to come to the hospital for treatment?"
	"Would you want to go to the ICU if needed to keep you alive?"
	"How would you feel about living in a nursing home when your condition worsens?"
Make a recommendation	"Given how important your independence is and your unwillingness to be cared for by others, I think that a do-not-resuscitate order would be most consistent with your goals, what do you think about that?"
	"Given that your biggest goal is to live to see your grandson graduate and your willingness to be hospitalized and cared for by others I recommend that your code status be full code"
Commit to f/u, set the stage to readdress concerns	"I'll bring this issue up again if you are hospitalized or when your condition worsens. I'm committed to helping to take care of you regardless of what you decide"

discussions about patient goals, values, hopes and fears. In states using the POLST paradigm the POLST form itself can be used to guide discussion of resuscitation as well as other treatment preferences. Resuscitation discussions also provide an excellent context in which to discuss patient preferences for level or location of care as well as to explore any undesired states that the patient would find an unacceptable quality of life.

Discussing Goals of Care When Things Are Not Going Well

In the absence of sudden cardiac death, chronic heart failure patients usually face a slow functional decline punctuated by disease exacerbations involving increasing symptoms and possibly repeated hospitalizations. When a patient's condition worsens it is important to again explore their understanding, clarify their goals and discuss prognosis in the setting of available treatment options [16]. Table 2.5 contains

Table 2.5 Discussing goals of care when things are not going well

Task (tool)	Suggested language
Assessing understanding (*Ask*)	"What is your sense of how things are going?"
Warning (*Tell*)	"I got the test results and they were not what we were hoping for."
Describing the "Big Picture" (Tell)	"He is very sick, we're worried he may not get any better."
Exploring changing wishes	"The care you are getting now is more aggressive than what you had previously said you wanted—how do you feel about that?"
	"Would you want to go back to the ICU if that is needed to keep you alive?"
Empathizing	"I wish we could fix this."
Re-Addressing levels of intervention, changing prognosis	"With what has happened now your chance of meaningful recovery if you were resuscitated would be very low."
Support	"I am so glad he has you to help him through this, it means a lot."
Clarifying next steps (*Ask*)	"So, we will continue this treatment for another 24 h then meet again"
	"Please tell us any questions you have now or going forward"

sample language for discussing goals of care in a patient whose condition is worsening. These discussions involve aspects of delivering bad or serious news, discussion of prognosis and readdressing resuscitation preferences. All of these areas also frequently involve responding to strong emotion.

Responding to Strong Emotion

During discussion of any of the above topics it is common for patients, their families and even healthcare providers to experience strong emotions. In response to threats, real or perceived, individuals commonly experience strong emotions, essentially a fight-or-flight response. In this setting the brain is not able to effectively deal with cognitive input. What can be confusing is that many people will express these strong emotions with what appear to be cognitive requests or conversations. Clinicians frequently respond to these requests with more information-giving, not realizing that the patient or family member is expressing strong emotions and not able to effectively receive cognitive information. Responding to strong emotion with empathetic statements is an effective method to help patients or families to understand their emotions and adjust to the information already received [13]. This will often reduce the emotional level and allow further discussion to proceed. Several examples of empathetic responses to statements with overt or underlying strong emotion are presented in Table 2.6.

Table 2.6 Responding to emotion

Sample patient/family statement	Possible response
"You said this treatment would help, why didn't it work?"	"I wish it had worked, I know you were hoping it would."
"Can't they just give him more medication, the medication always worked before?"	"It must be hard to see him so sick when he's pulled through this so many times before"
"How long will I live?"	"I'll answer all your questions the best I can. I wonder if it is frightening for you not knowing what will happen next or when"
"We have to do everything"	"What does doing everything mean to you?"
"You can't just give up on me"	"I won't. What do you mean by giving up?"
"Is he going to die?"	"I wish that were not true. What do you think would be most important to him if time were limited?"

Newer Models of Advance Care Planning

Respecting Choices is an advance care planning program originally developed in La Crosse, Wisconsin which has been used and studied widely [17]. This program involves training of ACP facilitators, education of healthcare providers and patients and integration of ACP into ongoing clinical care. The Respecting Choices program has been implemented in many communities in the US and abroad and has been demonstrated to improve adherence to patient preferences for treatment and to increase the use of ACP & AD [18, 19]. Respecting Choices was also implemented in a disease specific ACP intervention for patients with heart failure. In this case use of Respecting Choices was associated with a higher rate of AD completion as well as a higher use of hospice in those who died during the study period [20]. Respecting Choices has also been demonstrated to improve congruence between patient and surrogate reported goals and preferences [21, 22].

The POLST Paradigm was originated in Oregon and intended for use with patients with serious chronic illnesses [23]. POLST (Physician Order for Life-Sustaining Treatment) is organized around a brightly colored physician order form which is kept in the patient's home or in the chart if the patient is in a supervised care or living setting. The POLST requires a conversation between the patient (or his/her designated surrogate) and a healthcare practitioner. Completion of the form involves addressing resuscitation preference, preference of level of medical care (comfort measures only/limited treatments/full treatments) and preferences for the use of artificial nutrition/hydration. The POLST form is intended to travel with the patient across all care or living settings and is a legal physician's order, which is followed by medical professionals across sites. Rather than a stand-alone AD, the POLST is an ACP tool that can be used with or without existing directives. Use of the POLST has been demonstrated to improve documentation of patient preferences and adherence to those preferences in clinical care. In a nursing home population, with 'do-not-resuscitate' and 'comfort measures only' on a POLST none of the patients received unwanted intensive care, ventilator support or cardiopulmonary resuscitation [24]. Though not a decision-making tool per se, the POLST may be used as a guide for ACP discussions and represents a promising development in allowing patients with advanced illness to specify their preferences for care in a portable, legally binding form. The POLST paradigm has spread nationally; as of mid-2013, 43 states have a POLST-type program in place or in development [25].

Another promising advance care planning program is the Five Wishes paradigm. This guides patients through decision-making and documentation of "(1) The person I want to make care decisions for me when I can't; (2) The kind of medical treatment I want or don't want; (3) How comfortable I want to be; (4) How I want people to treat me; (5) What I want my loved ones to know" [26, 27]. Five Wishes is a legal advance directive document in many states.

The Role of Legal Documentation in Advance Care Planning

While ongoing communication with patients and their families is the preferred method of advance care planning, there are certain instances when complete documentation is very important. These include when there is disagreement between family members or when the patient chooses a surrogate decision-maker who is not the legal default decision-maker (unmarried partner, friend, etc.). It is also important for residents of nursing homes or other supervised living/care situations to have their preferences documented—this should include at a minimum designation of a surrogate decision-maker, resuscitation preferences, preferences for transfer to the hospital for acute illness and/or designation of comfort measure only when appropriate. In states where it is available the POLST form serves this purpose very well.

Summary

Advance healthcare directives were born as legal tools intended to protect patients from unwanted medical interventions. This role has evolved through time such that these directives form one piece of a larger process called "advance care planning" or "preparedness planning." Rather than simply a legal documentation process, this sort of planning represents a continuing conversation between providers, patients and their families. Such ongoing conversations support patient autonomy as well as improving the likelihood that patients will receive care consistent with their wishes at the end of life. All providers caring for patients with advanced heart failure or other cardiac disease should develop a repertoire of communication skills and approaches that support shared decision-making. At the level of healthcare organizations there is a need for continuing education and institutional support for addressing patient values, wishes and goals of care and in competently addressing cultural and spiritual issues at the end of life. At the societal level there is a growing need to support these important conversations as critical facets of caring for patients with chronic, life-limiting illnesses.

References

1. Kutner L. Due process of euthanasia: the living will, a proposal. Indiana Law J. 1969;44:539–54.
2. Sabatino CP. The evolution of health care advance planning law and policy. Milbank Q. 2010;88(2):211–39.
3. Hickman SE, Hammes BJ, Moss AH, Tolle SW. Hope for the future: achieving the original intent of advance directives. Hastings Cent Rep. 2005;S26–30.
4. Tulsky JA. Beyond advance directives: importance of communication skills at the end of life. JAMA. 2005;294(3):359–65.
5. Perkins HS. Controlling death: the false promise of advance directives. Ann Intern Med. 2007;147(1):51–7.
6. Fried TR, O'Leary J, Van Ness P, Fraenkel L. Inconsistency over time in the preferences of older persons with advanced illness for life-sustaining treatment. J Am Geriatr Soc. 2007;55(7):1007–14.
7. Institute of Medicine; Committee on Quality of Health Care in America. Crossing the quality chasm: a new health system for the 21st century. Washington, DC: National Academies Press; 2001.
8. Harding R, Selman L, Beynon T, Hodson F, Coady E, Read C, et al. Meeting the communication and information needs of chronic heart failure patients. J Pain Symptom Manage. 2008; 36(2):149–56.
9. Krakauer EL, Crenner C, Fox K. Barriers to optimum end-of-life care for minority patients. J Am Geriatr Soc. 2002;50(1):182–90.
10. Ho KK, Anderson KM, Kannel WB, Grossman W, Levy D. Survival after the onset of congestive heart failure in Framingham Heart Study subjects. Circulation. 1993;88(1):107–15.
11. Stewart S, MacIntyre K, Hole DJ, Capewell S, McMurray JJ. More 'malignant' than cancer? Five-year survival following a first admission for heart failure. Eur J Heart Fail. 2001;3(3):315–22.
12. Ahluwalia SC, Levin JR, Lorenz KA, Gordon HS. Missed opportunities for advance care planning communication during outpatient clinic visits. J Gen Intern Med. 2012;27(4):445–51.
13. Back A, Arnold RM, Tulsky JA. Mastering communication with seriously ill patients: balancing honesty with empathy and hope. Cambridge, UK/New York: Cambridge University Press; 2009. x, 158 p.
14. Goodlin SJ, Quill TE, Arnold RM. Communication and decision-making about prognosis in heart failure care. J Card Fail. 2008;14(2):106–13.
15. Loertscher L, Reed DA, Bannon MP, Mueller PS. Cardiopulmonary resuscitation and do-not-resuscitate orders: a guide for clinicians. Am J Med. 2010;123(1):4–9.
16. Goodlin SJ. Palliative care in congestive heart failure. J Am Coll Cardiol. 2009;54(5):386–96.
17. Romer AL, Hammes BJ. Communication, trust, and making choices: advance care planning four years on. J Palliat Med. 2004;7(2):335–40.
18. Detering KM, Hancock AD, Reade MC, Silvester W. The impact of advance care planning on end of life care in elderly patients: randomised controlled trial. BMJ. 2010;340:c1345.
19. Hammes BJ, Rooney BL, Gundrum JD. A comparative, retrospective, observational study of the prevalence, availability, and specificity of advance care plans in a county that implemented an advance care planning microsystem. J Am Geriatr Soc. 2010;58(7):1249–55.
20. Schellinger S, Sidebottom A, Briggs L. Disease specific advance care planning for heart failure patients: implementation in a large health system. J Palliat Med. 2011;14(11):1224–30.

21. Schwartz CE, Wheeler HB, Hammes B, Basque N, Edmunds J, Reed G, et al. Early intervention in planning end-of-life care with ambulatory geriatric patients: results of a pilot trial. Arch Intern Med. 2002;162(14):1611–8.
22. Briggs LA, Kirchhoff KT, Hammes BJ, Song MK, Colvin ER. Patient-centered advance care planning in special patient populations: a pilot study. J Prof Nurs. 2004;20(1):47–58.
23. Tolle SW, Tilden VP. Changing end-of-life planning: the Oregon experience. J Palliat Med. 2002;5(2):311–7.
24. Tolle SW, Tilden VP, Nelson CA, Dunn PM. A prospective study of the efficacy of the physician order form for life-sustaining treatment. J Am Geriatr Soc. 1998;46(9):1097–102.
25. Website NPTF. National POLST Task Force Website. http://www.polst.org2013. Accessed 12 Aug 2013.
26. Five Wishes. http://www.agingwithdignity.org. Last accessed Aug 2013.
27. Dyar S, Lesperance M, Shannon R, Sloan J, Colon-Otero G. A nurse practitioner directed intervention improves the quality of life of patients with metastatic cancer: results of a randomized pilot study. J Palliat Med. 2012;15(8):890–5.

Chapter 3
Palliative Care and Hospice in Patients with Advanced Cardiovascular Disease

Maria Dans and Kathleen Garcia

Abstract Advanced cardiovascular disease is common and frequently carries a high symptom burden. With its focus on symptom control, advance care planning, and family support, palliative care augments traditional approaches to the care of patients who are dying and their families. Cardiovascular disease is a leading cause of death; however, hospice services remain relatively underused among this patient population. Difficulties with prognostication in cardiovascular disease have contributed to this, along with cultural and practice differences between cardiology and hospice and palliative care. Improving end-of-life care for patients with cardiovascular disease will require a proactive and collaborative approach between practitioners of both specialties.

Keywords Advanced heart failure • Palliative care • Hospice • End-of-life

Abbreviations

AD	Advanced directive
CHD	Congenital heart disease
COPD	Chronic obstructive pulmonary disease
CPAP	Continuous positive airway pressure
DNR	Do not resuscitate
EOL	End of life
HF	Heart failure
ICD	Implantable cardioverter-defibrillators

M. Dans, MD (✉)
Department of Medicine, Washington University School of Medicine, St. Louis, MO, USA

Palliative Care Service, Internal Medicine 1, Barnes-Jewish Hospital,
4105 Queeny Tower, 1 Barnes-Jewish Hospital Plaza, St. Louis, MO 63110-1093, USA
e-mail: mdans@dom.wustl.edu

K. Garcia, MD
Department of Medicine, Barnes-Jewish Hospital, St. Louis, MO, USA

S.J. Goodlin, M.W. Rich (eds.), *End-of-Life Care in Cardiovascular Disease*,
DOI 10.1007/978-1-4471-6521-7_3

ICU	Intensive care unit
NYHA	New York Heart Association
SSRI	Selective serotonin reuptake inhibitor
VAD	Ventricular assist device

Key Points

- Palliative care and aggressive symptom management should be provided for all HF patients regardless of the stage of disease.
- Because the course of HF can be variable it is important to address EOL issues early in the disease course. However, it is also important to recognize that patients and providers may be reluctant to discuss death and dying, especially early in the disease. Discussions of preferences for EOL care should be an ongoing process rather than a single conversation.
- The mainstay of symptom management in advanced HF is treatment of the underlying disease. Other comorbidities such as sleep-disordered breathing, chronic pain, and depression, can contribute to symptoms and should be screened for and addressed.
- Technologically advanced HF therapies such as VADs and ICDs, may result in dramatic improvements in patients' quality of life and/or survival, but can also present challenges for EOL care. Anticipation and preparation for these challenges should begin before implantation.

Introduction

Over the course of the past century, how and where people die in the US has changed dramatically. Due to tremendous advances in public health infrastructure and medical treatments, people are living longer and healthier lives. Along with this, an increasing number of people are also facing challenges at the end of their lives posed by chronic, progressive, and incurable illnesses. Nowhere is this more evident than in the treatment of cardiovascular disease. Almost 600,000 people die each year from heart disease in the US, yet few of those people receive hospice or other systematic end-of-life (EOL) care. Heart disease is the leading cause of death in the US, but it is the diagnosis for only 11.2 % of hospice patients [1]. This chapter will address issues pertaining to the provision of palliative and hospice care for patients with heart disease; we will focus on considerations for patients with advanced heart failure (HF), as this disease represents a final common pathway for many forms of cardiac disease. Furthermore, many issues faced by HF patients at EOL are similar to those faced by people with other forms of cardiac disease, such as valvular disorders, persistent arrhythmias, and coronary artery disease. An estimated 6.6 million adults in America suffer from HF, and the prevalence of this disease is expected to grow by 25 % over the next 20 years [2]. The cost of care for HF patients is significant; in 2010 the total was estimated at $40 billion [3]. Patients with advanced HF carry a heavy symptom burden, which can lead to decreased functional capability, social isolation, and depression [4]. Caregivers of patients with advanced HF may also experience deterioration in their own health and quality of life [5]. Symptom palliation in advanced HF has not been well-studied; most recommendations are extrapolated from studies of symptom management in other diseases, such as chronic obstructive pulmonary disease (COPD) and cancer. Despite improvements in advanced HF therapy, the disease remains incurable and progressive for most patients. Many patients and health care providers, however, have difficulty recognizing it as a "fatal disease" [6]. This is, in part, due to the variable course of HF which increases uncertainty about when to initiate EOL care. Growing recognition of the importance of providing coordinated EOL care to HF patients has led to calls for earlier incorporation of palliative care into the care of HF patients, including changes in major cardiology society guidelines [7, 8]. However, providing EOL care remains challenging. Palliative care and hospice practitioners have also realized that the traditional cancer-focused paradigm for EOL care must be expanded to provide a better fit for HF patients and their families [9].

Definition of Hospice and Palliative Care

Palliative care focuses on enhancing quality of life for patients with serious illness and their families by encouraging advance care planning and providing aggressive control of symptoms. It may be provided concurrently with life-prolonging therapies regardless of the stage of the disease. Since illness affects people's lives on many levels, palliative care practitioners use an interdisciplinary team approach to expand traditional disease-model medical treatments. Holistic care of chronically ill patients must address not simply the physical aspects of a chronic disease, but the psychological, emotional, spiritual, social, and financial sequelae of illness as well. Palliative care frequently occurs in an inpatient consultative setting, but may also be provided in outpatient clinics or via home-based care. Hospice is a specific type of palliative care designed for terminally-ill patients who have a life expectancy of less than 6 months. Hospice agencies provide team-based services to patients, families, and caregivers in the home or an institution.

Integration of Palliative Care and Hospice in the Treatment of Patients with End Stage Cardiovascular Disease

The uncertain disease trajectory of HF is only one obstacle to timely integration of palliative and hospice care into the treatment of advanced HF patients. Improvements in HF treatment options, a lack of published evidence to support palliative care interventions, and cultural and practice differences between cardiologists and

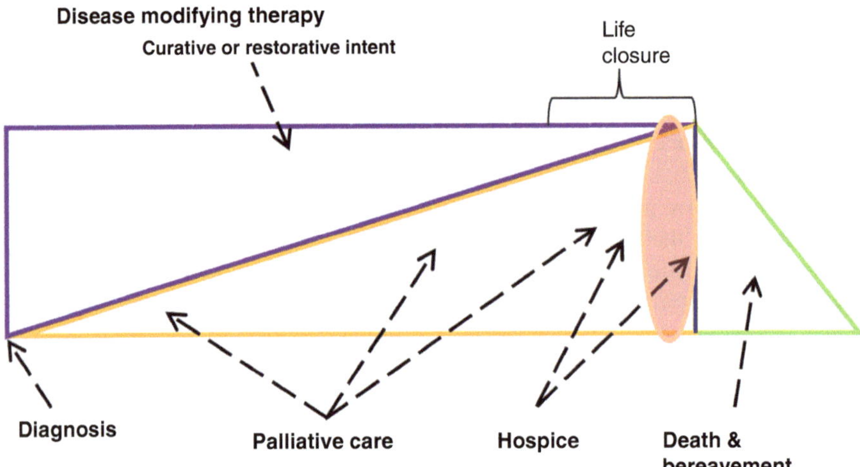

Fig. 3.1 Traditional model for the integration of palliative care and hospice in a terminal illness. Note that this paradigm may be more applicable to cancer patients than patients with advanced cardiovascular disease, and may contribute to the late initiation of discussions surrounding end-of-life care (*ellipse*) (Adapted from Goldstein [9]. Used with permission)

palliative care and hospice specialists also complicate the picture. Still, palliative interventions, including hospice, can be very beneficial for patients with advanced HF. Several major cardiology societies, including the American College of Cardiology, recommend advance care planning and consideration of hospice and palliative care services for HF patients at EOL [8] (Fig. 3.1).

Estimation of Prognosis

Progression of disease in advanced HF is characterized by acute and severe exacerbations, requiring aggressive medical management. Treatment often provides rapid and durable improvement in symptoms and functional status, but advanced HF patients remain susceptible to rapid decompensation and sudden death. Several risk-profiling scores have been developed to improve physician estimates of prognosis; even so, it remains difficult to predict the disease trajectory for many patients with advanced HF [10–13]. The Seattle Heart Failure Model uses 24 variables to predict 1–3- year survival among patients aged 85 or younger, although it may overestimate survival, particularly among the elderly [13, 14]. Simpler models using fewer variables also may be helpful [15]. Seminal events, such as re-hospitalizations, may provide additional predictors of mortality; median survival after the fourth hospitalization is 1 year or less [16]. Adding to the complexity, many patients with advanced HF have multiple other comorbidities [17]. Under these circumstances, determining prognosis can be difficult not only for cardiologists but also for palliative care and hospice practitioners who may be accustomed to the more predictable clinical decline at EOL seen in patients with solid organ tumors [18]. Prognostic uncertainty in advanced HF is a problem that is unlikely to be resolved; physicians may miss opportunities to explore patient wishes regarding EOL care unless discussions of prognosis and the uncertainty surrounding it begin early in the course of therapy.

Communication Around Prognosis

Physicians may feel reluctant to address the possibility of dying with their patients, but patients and their caregivers frequently find such information helpful [19]. Because of the uncertainty intrinsic to advanced HF prognosis, discussions that include acknowledgement of HF as a life-limiting illness should begin early in the course of therapy. Communication about prognosis and preferences regarding EOL care should be viewed as an ongoing process. Initial discussions may be quite brief, but as the disease progresses, more in-depth conversations will become necessary. To communicate effectively in these difficult situations, it is important to assess what the patient knows and wants to know about the disease and its prognosis. Cognitive impairment is common among HF patients [20]; if appropriate, clinicians should offer to include caregivers in the discussion [21]. One simple but effective model for "bad news" conversations is "ask-tell-ask" [22]. The clinician starts by *asking* about the patient's understanding, then *tells* information, and finally *asks* for feedback from the patient. Eliciting questions and establishing a common understanding require careful attention not just to the words spoken but also to the emotions displayed. The extent of discussion will depend on the readiness of the patient and family to hear and accept what the clinician is saying. In the event that a patient is reluctant to address EOL issues, the clinician may conclude the discussion by asking permission to raise the topic again at a future visit.

Timing of Palliative Care and Hospice Referrals

Palliative care referral is appropriate if a HF patient could benefit from specific attention to symptom management and help with advanced care planning, regardless of whether the patient is receiving aggressive medical management. In particular, palliative care consultation may benefit patients who are undergoing evaluation for advanced HF therapies, such as continuous inotrope infusions, ventricular assist devices (VAD), and heart transplantation [23]. Other prompts for palliative care consultations may include worsening HF requiring repeated hospitalizations, and patient or family request for more information. Not all palliative care, however, needs to be provided by specialists; in some hospitals, checklists are used to identify patients in need of "primary" or basic palliative care interventions versus those in need of more complex palliative care services [24]. Integration of palliative care concepts into HF clinics may allow for gentler transitions when patients with advanced heart disease approach EOL.

Although HF patients comprise only a small portion of hospice beneficiaries, use of hospice services has been increasing both in general and in the HF population in recent years [25]. For any given disease, the Medicare hospice benefit requires a physician to certify that a patient's life expectancy is 6 months or less if the disease were to run its natural course. In the case of HF, the Medicare guidelines rely heavily on the New York Heart Association (NYHA) symptom class and associated comorbidities. One obstacle to timely hospice referral for HF patients is that even using the best risk profiling scores, it is difficult to give an estimate of prognosis that is precise enough to fit in the 6-month window. In addition, hospice agencies are not uniform in their coverage of advanced HF treatments, particularly more expensive therapies like home inotrope infusions and VAD therapy. It is important that the clinician, patient, and family be aware of potential limitations prior to initiation of advanced HF therapies, as these therapies may dictate where the patient can receive EOL care. Early integration of discussions about patient preferences for EOL care may help when it is time to transition to hospice care.

Advanced Care Planning

An advance directive (AD) is a legal document that provides guidance about a person's wishes for medical care. ADs frequently also designate a "durable power of attorney for healthcare" or surrogate decision-maker. These documents go into effect only if a person becomes unable to speak for himself. A recent study found that AD use is low among HF patients, but that patients who had completed an AD were less likely to be transferred to an intensive care unit (ICU) or to receive mechanical ventilation [26]. In addition, good patient-physician communication around advanced care planning improves patient satisfaction with their medical care [27]. Palliative care consultation has been shown to increase preparedness planning and use of ADs amongst patients being evaluated for VAD placement [28]. The idea that one should "hope for the best and prepare for the worst" will be recognized by most people. This concept may be used to communicate the clinician's desire to support the patient in their quest for life prolonging therapy in the face of a life limiting illness [29]. Focusing conversations about patient preferences for EOL care on "big picture goals" and values, instead of specific treatment decisions, may prove helpful in guiding future care [21].

Symptom Management

Patients with end-stage cardiovascular disease suffer from a significant burden of symptoms, due to both the pathophysiology of HF and to comorbidities common in HF patients, such as arthritis and sleep-disordered breathing. Although comorbid conditions must be addressed, the cornerstone of symptom management in advanced HF is optimization of the medical treatment itself. Common symptoms reported by patients with advanced HF include dyspnea, pain, fatigue, anorexia, depression, and anxiety [30].

Dyspnea

The first line of treatment for dyspnea in advanced HF is management of fluid status and systolic dysfunction with diuretics, vasodilators, intravenous inotropes, fluid restriction, and reduced sodium diet. As HF progresses, the ability to manage dyspnea by optimizing volume status may be hampered by hypotension and renal insufficiency. Although the mechanisms are not fully understood, low-dose opioids may provide relief in dyspnea refractory to diuresis [31]. Oxygen supplementation is frequently used to treat dyspnea, particularly in hypoxemic patients. Even patients without significant hypoxemia may derive benefit; however, it is not clear whether this is due to the oxygen itself or to the sensation of air flow [32]. Benzodiazepines also may be used as a second-line therapy, particularly for anxiety associated with dyspnea at EOL [33].

Pain

Pain is a common and debilitating symptom in patients with HF [34]. Although the causes are not completely understood, pain may result from the underlying pathophysiology of HF itself or it may be associated with frequently occurring comorbidities, including degenerative arthritis, anxiety, and depression [34, 35]. General principles of pain management can be applied to the treatment of pain in HF patients; however, special attention must be paid to drug selection and dosing, and therapeutic effectiveness must be frequently reassessed [36]. Non-steroidal anti-inflammatory agents are not recommended in HF patients due to their tendency to cause sodium and fluid retention; worsen kidney function; increase the risk of gastrointestinal bleeding; and potentiate adverse effects of important HF treatments, such as diuretics, angiotensin-converting-enzyme inhibitors, and angiotensin II receptor blockers [37].

Fatigue

Many end stage HF patients struggle with fatigue, which may be associated with low cardiac output, depression, obstructive sleep apnea, and anemia. Screening and treatment for underlying causes of fatigue can improve patients' quality of life.

Treatment of sleep apnea with continuous positive airway pressure (CPAP) has been shown to improve heart function, mental alertness and fatigue in patients with HF [37]. Decreased cardiac output may be treated with intravenous inotropes if it is consistent with the patient's goals of care. When appropriate, treatment of anemia with erythropoietin can improve exercise tolerance and overall quality of life in patients with advanced HF [38]. In addition, graded exercise programs, low-dose opioids, and caffeine have shown benefit in treating fatigue associated with exertion [39–41].

Depression and Anxiety

Patients with advanced HF face uncertainty about the future and loss of independence. Depression is common, but may be difficult to recognize in advanced HF, as many of the symptoms of depression are also symptoms of HF [42]. Supportive counseling may be beneficial, as may some complementary and alternative medicine interventions, such as mindfulness-based stress reduction [43]. For those patients who require pharmacologic treatment, selective serotonin reuptake inhibitors (SSRIs), tricyclic antidepressants, benzodiazepines, and psychostimulants, such as methylphenidate, have all shown some benefit in the treatment of depression, particularly when used in conjunction with counseling [44]. Benzodiazepines should be used with caution in the elderly due to potential adverse effects of sedation and confusion. Psychostimulants may be a better option in patients for whom a rapid response is important, although patients may experience worsening anxiety or agitation. All HF patients started on antidepressant medication should be monitored for mental status changes, prolongation of the QT interval, orthostatic hypotension, and hyponatremia.

Anorexia and Cachexia

Weight loss and anorexia in advanced HF patients are thought to be related to neurohormonal effects of HF itself. Little data exist about the effectiveness of pharmacological treatment of anorexia and cachexia, therefore potentially treatable underlying causes of anorexia must be addressed, including hypothyroidism, depression, intestinal edema, and uncontrolled pain [45]. Appetite stimulants may be helpful in cancer-related anorexia, but can cause side effects that may outweigh the benefits in HF patients. Discussions about the reasons for anorexia and weight loss in advanced HF may allow patients and families to focus less on the quantity of food consumed and more on the quality of the experience. Encouraging smaller servings of favorite foods and liberalizing dietary restrictions when appropriate can benefit the patient with end stage heart disease.

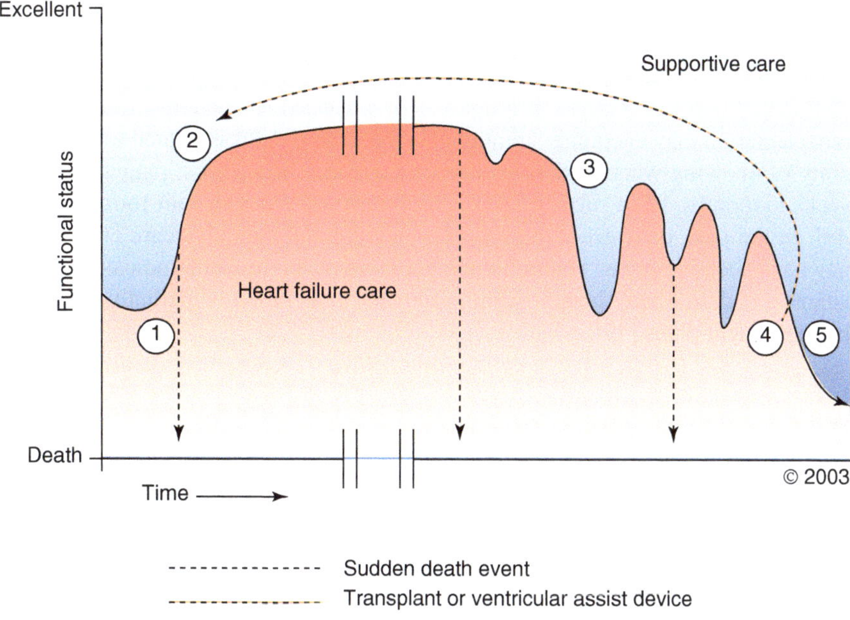

Fig. 3.2 Schematic course of Stage C and D heart failure. Sudden death may occur at any point along the course of illness. (*1*) Initial symptoms of heart failure (HF) develop and HF treatment is initiated. (*2*) Plateaus of variable length may be reached with initial medical management or after mechanical circulatory support (MCS) or heart transplant. (*3*) Functional status declines with variable slope, with intermittent exacerbations of HF that respond to rescue efforts. (*4*) Stage D HF, with refractory symptoms and limited function. Note that heart transplantation or MCS may result in a return to an earlier plateau on the disease trajectory. (*5*) End of life (Adapted with permission from Goodlin et al. [17])

Technological Advances and End-of-Life Care

Over the past several decades, there has been tremendous progress in the development of highly technological treatment modalities for advanced HF. Cardiac transplantation, the "gold standard" for the treatment of advanced HF, is limited by the number of donor hearts, but the number of patients receiving some form of device therapy has been rising steadily for the past 10 years. Although not all people with advanced HF receive these treatments, they deserve special mention because they can complicate EOL care Fig. 3.2.

Heart Transplantation

Despite improvements in medical and device therapy for HF, heart transplantation remains the best option for advanced HF patients who qualify for it. The primary limitation of heart transplantation is the scarcity of donor hearts. The annual number of heart transplants in the US is around 2,200, and there has been little change in this figure since 1998 [46]. Perioperative mortality rates have dropped since the 1980s; patients who survive their first year have a 63 % chance of surviving to 10 years [47]. In addition to the physical symptoms of advanced HF, patients awaiting heart transplantation may encounter emotional or existential distress related to the uncertainty of knowing whether they will receive a donor heart. Most patients wait for weeks to months, but a substantial proportion (30–37 %) may wait for over a year [48]. In the past, concurrent transplant evaluation and palliative care consultation may have been viewed as contradictory, but more recent research indicates that this patient population may benefit from systematic integration of palliative care into their treatment plans [49].

Mechanical Circulatory Support

Ventricular assist devices (VADs) have been shown to decrease morbidity and mortality among patients with HF refractory to medical management [50]. As a result, VADs have become an important HF treatment modality, both as a bridge to cardiac transplantation and as "destination therapy." From 2006 to 2010, the number of VADs implanted in the US increased from 206 to 1,451 [51]. Mean survival for advanced HF patients increases from 6 months with medical therapy alone to 2 years with the addition of continuous-flow VADs [52–54]. This technology is expensive – the cost is almost six times higher than the cost of medical management; however, continuous-flow VAD patients may experience dramatic improvements in quality of life compared to medically managed patients [55].

Despite advancements in VAD technology, adverse events may occur, leading to increased morbidity, hospitalization, and death. The most common causes of death among VAD patients include stroke, worsening HF or VAD failure, sepsis, multiorgan failure, and infection [56–60]. EOL care for patients with VADs may be complicated by the need to obtain care at tertiary care centers as smaller hospitals lack experience with the devices. In addition, many hospice organizations may be unable to support the care of VAD patients due to lack of familiarity and high cost. As the VAD patient population grows, questions of how to handle death and dying in these patients and the need for formalized support for patients, caregivers, and staff will become more prominent.

Pacemakers and Implantable Cardioverter-Defibrillators

Pacemakers have improved the quality and length of life for patients with bradyarrhythmias for decades [61]. More recently, implantable cardioverter-defibrillators (ICDs) have demonstrated substantial survival benefits for patients at risk for sudden cardiac death [62]. Cardiac resychronization therapy may also improve symptoms and survival [63]. Since a major benefit of cardiac pacing therapies is symptom relief, the question of how to manage these devices at EOL is not usually problematic. The majority of ICDs, on the other hand, are implanted not for symptom relief, but to reduce the risk of sudden death. Up to 20 % of patients with ICDs can experience painful shocks in the last few weeks of their lives, and these shocks can impair patients' quality of life and cause distress amongst their caregivers [64]. Patients and caregivers may also have misconceptions about what ICDs do, and what to expect if they are turned off [65]. Guidelines and consensus statements issued by the major cardiology societies strongly encourage physicians and patients to discuss the option of ICD deactivation at the EOL prior to implantation, and it is recommended that organizations caring for patients with ICDs have a formal policy regarding ICD deactivation [66]. Unfortunately many hospice programs do not have such policies – a 2010 survey of 900 hospices across the US revealed that patients in hospices that had established policies regarding ICD deactivation were twice as likely to have their device turned off, but that only 10 % of hospices had such policies [67]. Although deactivating an ICD may shorten survival time, physicians should emphasize that it will not result in immediate death and that the process of deactivation is not painful. If a patient decides that deactivation is the right choice, the ICD function may be deactivated by reprogramming the device, which can occur at home, in the clinic, or in the hospital. Organizations that care for patients with ICDs should have protocols in place for patients who are actively dying but have not deactivated their ICD. The ICD function of most devices can be disabled by placing a magnet on top of the device generator; as long as the magnet remains in place, the ICD will not sense arrhythmias and will not deliver shocks (pacemaker functions will not be disabled by the magnet, although they may revert to the default settings).

IV Intravenous Inotropes

Continuous infusions of inotropes have not demonstrated a survival benefit for advanced HF patients; however, in some cases these medications may improve symptoms. The role of continuous IV inotropes in the EOL care of advanced HF patients has not been well-studied. Advanced HF patients who do not qualify for transplant or VAD treatment and are started on palliative inotrope infusions have very high mortality rates and may have a shortened lifespan as a result of inotropic therapy [68]. There is no easy way to know whether a person facing EOL will choose quality of life or quantity of life. Talking to patients about options before they reach the terminal stages of disease is still the best way to understand treatment preferences [69].

Cardiopulmonary Resuscitation and Do-Not-Resuscitate Orders

Hospitalized patients with advanced HF are far less likely to have Do-Not-Resuscitate (DNR) orders in place than patients with advanced cancer; they are also far more likely to receive life-sustaining treatments during their final hospitalization [70, 71]. Some of the reluctance on the part of both physicians and patients to discuss or institute DNR orders may be a perception that patients with DNR orders get substandard care. Indeed, there is some evidence that advanced HF patients who have DNR orders are less likely to meet quality of care performance measures [72]. Providers should be explicit that DNR orders apply only to resuscitation, and that optimal management of HF and its symptoms is the goal for all patients, regardless of resuscitation preferences. Inquiring about "code status" is not the best way to introduce a discussion of EOL issues, but this is commonly how the subject is broached in acute care hospitals [73]. A more fruitful way of framing the question might be to focus not on what people want when they die (or almost die), but how they want to live up to that point. For some patients with advanced HF, the decision to request DNR status may represent a willingness to trade length of life for quality of life [74]. Considered discussion of the patient's goals and of how various treatment options will promote the best quality of life should begin in the outpatient setting and continue in the inpatient setting. If CPR will help the patient achieve his or her goals, it should be offered for consideration; if not, it should be presented as "inadvisable" [75].

Care of Adults with Congenital Heart Disease

Although congenital heart disease (CHD) affects relatively few people, it is one of the most common birth defects. In the past, infant mortality from CHD was quite high, but advances in cardiology and cardiac surgery have changed the demographics of this disease; over 90 % of children born with CHD now survive to adulthood [76]. Further study is needed to develop guidelines for EOL care of this patient population, but strategies of care used for other patient populations with early onset disease, such as cystic fibrosis and childhood cancers, may prove useful [77]. Because symptoms of worsening HF may not manifest until late in the disease course, these patients require an early and proactive approach to discussions about life-prolonging treatment options and wishes regarding EOL care [78].

Hospice Care and Advanced Heart Failure Treatments

In part because of difficulties with prognostication, use of hospice care remains low among patients with advanced HF relative to patients with cancer [79]. In addition, clinicians and advanced HF patients may have misconceptions about hospice, believing that it is only for people who are actively dying and not for people who maintain some capacity to function. Patients with HF are much more likely than patients with cancer to be referred to hospice during an acute hospitalization, and they are also much more likely to return to the hospital despite hospice enrollment [80]. Hospices are reimbursed on a *per diem* basis – currently most US hospices receive about $150/day to cover all of a patient's care, including personnel, medications, and durable medical equipment [81]. As a result, hospices generally provide oral medications only; many are unable to support more high-technology treatments like continuous inotrope infusions, bi-level positive airway pressure (BiPAP), or VADs. In addition, some hospices may not be given, or do not ask for, necessary clinical data when admitting HF patients [82]. Dealing with the complex EOL care of patients with advanced HF will require specialists in cardiology and in hospice and palliative care to refine the current practice. Cardiology specialists should continue to be involved when advanced HF patients are referred to hospice. Educating hospice and palliative care practitioners about advanced HF remains a critical need, and will only be improved by a collaborative approach.

References

1. Center for Disease Control and Prevention. Fastats: heart disease. Available at: www.cdc.gov/nchs/fastats/heart.htm.
2. Roger VL, Go AS, Lloyd-Jones DM, Benjamin EJ, Berry JD, Borden WB, et al. Heart disease and stroke statistics – 2012 update: a report from the American Heart Association. Circulation. 2012;25:e2–220; originally published online Dec 15 2011.
3. Writing Group Members, Lloyd-Jones DM, Adams RJ, Brown TM, Carnethon M, Dai S, et al. Heart disease and stroke statistics – 2010 update: a report from the American Heart Association. Circulation. 2010;121:e46–215; originally published online Dec 17 2009.
4. Bekelman DB, Hutt E, Masoudi FA, Kutner JS, Rumsfeld JS. Defining the role of palliative care in older adults with heart failure. Int J Cardiol. 2008;125(2):183–90.
5. Hwang B, Fleischmann KE, Howie-Esquivel J, Stotts NA, Dracup K. Caregiving for patients with heart failure: impact on patients' families. Am J Crit Care. 2011;20(6):431–41.
6. Murray SA, Boyd K, Kendall M, Worth A, Benton TF, Calusen H. Dying of lung cancer or cardiac failure: prospective qualitative interview study of patients and their carers in the community. BMJ. 2002;325:929–34.
7. Yancy CW, Jessup M, Bozkurt B, Butler J, Casey Jr DE, Drazner MH, et al. 2013 ACCF/AHA guideline for the management of heart failure: a report of the American College of Cardiology Foundation/American Heart Association Task Force on Practice Guidelines. J Am Coll Cardiol. 2013;62(16):e147–239.
8. Heart Failure Society of America, Lindenfeld J, Albert NM, Boehmer JP, Collins SP, Ezekowitz JA, et al. HFSA 2010 comprehensive heart failure practice guideline. J Card Fail. 2010;16(6):e1–194.
9. Goldstein N. How to engage your heart failure colleagues: overcoming barriers, improving technique [MP3 and PDF files]. New York: Center to Advance Palliative Care; 2012.
10. Cristakis NA, Lamont EB. Extent and determinants of error in doctors' prognoses in terminally ill patients: a prospective cohort study. BMJ. 2000;320(7233):469–72.
11. The SUPPORT Principal Investigators. A controlled trial to improve care for seriously ill hospitalized patients. The study to understand prognoses and preferences for outcomes and risks of treatment (SUPPORT). JAMA. 1995;274(20):159–8.

12. Goldberg LR, Jessup M. A time to be born and a time to die. Circulation. 2007;116:360.
13. Levy WC, Mozaffarian D, Linker DT, Sudradhar SC, Anker SD, Cropp AB, et al. The Seattle heart failure model: prediction of survival in heart failure. Circulation. 2006;113(11):1424–33. Epub 2006 Mar 13.
14. Benbarkat H, Addetia K, Eisenberg MJ, Sheppard R, Filion KB, Michel C. Application of the Seattle heart failure model in patients >80 years of age enrolled in a tertiary care heart failure clinic. Am J Cardiol. 2012;110(11):1663–6.
15. Huynh BC, Rovner A, Rich MW. Identification of older patients with heart failure who may be candidates for hospice care: development of a simple four-item risk score. J Am Geriatr Soc. 2008;56(6):1111–5.
16. Setoguchi S, Stevenson LW, Schneeweiss S. Repeated hospitalizations predict mortality in the community population with heart failure. Am Heart J. 2007;154:260–6.
17. Goodlin SJ, Hauptman PJ, Arnold R, Grady K, Hershberger RE, Kutner J, et al. Consensus statement: palliative and supportive care in advanced heart failure. J Card Fail. 2004;10:200–9.
18. Maltoni M, Caraceni A, Brunelli C, Broeckaert B, Christakis N, Eychmueller S, et al. Prognostic factors in advanced cancer patients: evidence-based clinical recommendations – a study by the Steering Committee of the European Association for Palliative Care. J Clin Oncol. 2005;23(25):6240–8.
19. Emanuel EJ, Fairclough DL, Wolfe P, Emanuel LL. Talking with terminally ill patients and their caregivers about death, dying, and bereavement: is it stressful? Is it helpful? Arch Intern Med. 2004;164(18):1999–2004.
20. Hajduk AM, Kiefe CI, Person SD, Gore JG, Saczynski JS. Cognitive change in heart failure: a systematic review. Circ Cardiovasc Qual Outcomes. 2013;6(4):451–60.
21. Back A, Arnold R, Tulsky J. Mastering communication with seriously Ill patients. Cambridge: Cambridge University Press; 2009.
22. Goodlin SJ, Quill TE, Arnold RM. Communication and decision-making about prognosis in heart failure care. J Card Fail. 2008;14:106–13.
23. Swetz K, Freeman M, AbouEzzedine O, Carter KA, Boilson BA, Ottenberg AL, et al. Palliative medicine consultation for preparedness planning in patients receiving left ventricular assist devices as destination therapy. Mayo Clin Proc. 2011;86(6):493–500.
24. Weissman DE, Meier DE. Identifying patients in need of a palliative care assessment in the hospital setting: a consensus report from the Center to Advance Palliative Care. J Palliat Med. 2011;14(1):1–7.
25. Unroe KT, Greiner MA, Hernandes AF, Whellan DJ, Kaul P, Schulman KA, et al. Resource use in the last 6 months of life among Medicare beneficiaries with heart failure, 2000–2007. Arch Intern Med. 2011;171:196–203.
26. Dunlay SM, Swetz KM, Mueller PS, Roger VL. Advance directives in community patients with heart failure. Circ Cardiovasc Qual Outcomes. 2012;5:283–9.
27. Tierney WM, Dexter PR, Gramelspacher GP, Perkins AJ, Zhou XH, Wollensky FD. The effect of discussions about advance directives patients: satisfaction with primary care. J Gen Intern Med. 2001;16:32–40.
28. Swetz KM, Mueller PS, Ottenberg AL, Dib C, Freemand MR, Sulmasy DP. The use of advance directives among patients with left ventricular assist devices. Hosp Pract. 2010;39:78–84.
29. Back AL, Arnold RM, Quill TTE. Hope for the best and prepare for the worst. Ann Intern Med. 2003;138:439.
30. Blinderman CD, Homel P, Billings JA, Portenoy RK, Tennstedt SL. Symptom distress and quality of life in patients with advanced congestive heart failure. J Pain Symptom Manage. 2008;35(6):594–603.
31. Johnson MJ, Abernathy AP, Currow DC. Gaps in the evidence base for refractory breathlessness: a future work plan? J Pain Symptom Manage. 2011;43:614–24.
32. Johnson MJ, Abernathy AP, Currow DC. The evidence base for oxygen for chronic refractory breathlessness: issues, gaps, and a future work plan. J Pain Symptom Manage. 2013;45:763–75.

33. Simon ST, Higginson IJ, Booth S, Harding R, Bausewein C. Benzodiazepines for relief of breathlessness in advanced malignant and non-malignant diseases in adults. Cochrane Database Syst Rev. 2010;(1):CD007354.
34. Goodlin SJ, Wingate S, Albert NM, Pressler SJ, Houser J, Kwon J, et al. Investigating pain in heart failure patients: the pain assessment, incidence, and nature in heart failure (PAIN_HF) study. J Card Fail. 2012;18:776–83.
35. Levenson JW, McCarthy EP, Lynn J, Davis RB, Phillips RS. The last six months of life for patients with congestive heart failure. J Am Geriatr Soc. 2000;48(5 Suppl):S101–9.
36. Goodlin SJ, Light-McGroary KA. The challenges of understanding and managing pain in the heart failure patient. Curr Opin Support Palliat Care. 2013;7(1):14–20.
37. Page J, Henry D. Consumption of NSAIDs and the development of congestive heart failure in elderly patients: an underrecognized public health problem. Arch Intern Med. 2000;160:777–84.
38. Kotecha D, Ngo K, Walters JA, Manzano L, Palazzuoli A, Flather MD. Erythropoietin as a treatment of anemia in heart failure: systematic review of randomized trials. Am Heart J. 2011;161:822–31.
39. van der Meer S, Zwerink M, van Brussel M, van der Valk P, Wajon E, van der Palen J. Effect of outpatient exercise training programmes in patients with chronic heart failure: a systematic review. Eur J Prev Cardiol. 2012;19(4):795–803.
40. Williams SG, Wright DJ, Marshall P, Reese A, Tzeng BH, Coats AJ, Tan LB. Safety and potential benefits of low dose diamorphine during exercise in patients with chronic heart failure. Heart. 2003;89(9):1085–6.
41. Notarius CF, Morris B, Floras JS. Caffeine prolongs exercise duration in heart failure. J Card Fail. 2006;12(3):220–6.
42. Rutledge T, Reis VA, Linke SE, Greenberg BH, Mills P. Depression in heart failure: a meta-analytic review of prevalence, intervention effects, and associations with clinical outcomes. J Am Coll Cardiol. 2006;48:1527–37.
43. Woltz PC, Chapa DW, Friedmann E, Son H, Akintade B, Thomas SA. Effects of interventions on depression in heart failure: a systematic review. Heart Lung. 2012;41(5):469–83.
44. Moudgil R, Haddad H. Depression in heart failure. Curr Opin Cardiol. 2013;28(2):249–58.
45. Fudim M, Wagman G, Altschul R, Yucel E, Bloom M, Vittorio TJ. Pathophysiology and treatment options for cardiac anorexia. Curr Heart Fail Rep. 2011;8(2):147–53.
46. Vega JD, Moore J, Murray S, Chen JM, Johnson MR, Dyke DB. Heart transplantation in the United States, 1998–2007. Am J Transplant. 2009;9(Part 2):932–41.
47. Stehlik J, Edwards LB, Kucheryavaya AY, Benden C, Christie JD, Dipchand AI, et al. Registry of the International Society for Heart and Lung Transplantation: twenty-ninth official adult heart transplant report – 2012. J Heart Lung Transplant. 2012;31(10):1052–64.
48. Johnson MR, Meyer KH, Haft J, Kinder D, Webber SA, Dyke DB. Heart transplantation in the United States, 1999–2008. Am J Transplant. 2010;10(Part 2):1035–46.
49. Schwartz ER, Baraghoush A, Morrissey RP, Shah AB, Shinde AM, Phan A, et al. Pilot study of palliative care consultation in patients with advanced heart failure referred for cardiac transplantation. J Palliat Med. 2011;15(1):12–5.
50. Miller LW. Controversies in cardiovascular medicine: left ventricular assist devices are underutilized. Circulation. 2011;123:1552–8.
51. Birks EJ. A changing trend toward destination therapy: are we treating the same patients differently? Tex Heart Inst J. 2011;38(5):552–4.
52. Fang JC. Rise of the machines — left ventricular assist devices as permanent therapy for advanced heart failure. N Engl J Med. 2009;361:2282–5.
53. Rose EA, Gelijns AC, Moskowitz AJ, Heitjan DF, Stevenson LW, Dembitsky W, Randomized Evaluation of Mechanical Assistance for the Treatment of Congestive Heart Failure (REMATCH) Study Group, et al. Long-term use of a left ventricular assist device for end-stage heart failure. N Engl J Med. 2001;345(20):1435–43.
54. Starling RC, Naka Y, Boyle AJ, Gonzalez-Stawinski G, John R, Jorde U, et al. Results of the post-U.S. Food and Drug Administration-approval study with a continuous flow left ventricular

assist device as a bridge to heart transplantation: a prospective study using the INTERMACS (Interagency Registry for Mechanically Assisted Circulatory Support). J Am Coll Cardiol. 2011;57(19):1890–8.
55. Rogers JG, Bostic RR, Tong KB, Adamson R, Russo M, Slaughter MS. Cost-effectiveness analysis of continuous-flow left ventricular assist devices as destination therapy. Circ Heart Fail. 2012;5:10–6.
56. Lietz K, Long JW, Kfoury AG, Slaughter MS, Silver MA, Milano CA, et al. Outcomes of left ventricular assist device implantation as destination therapy in the post-REMATCH era: implications for patient selection. Circulation. 2007;116(5):497–505.
57. Holman WL, Pae WE, Teutenberg JJ, Acker MA, Naftel DC, Sun BC, et al. INTERMACS: interval analysis of registry data. J Am Coll Surg. 2009;208(5):755–61.
58. Brush S, Budge D, Alharethi R, McCormick AJ, MacPherson JE, Reid BB, et al. End-of-life decision making and implementation in recipients of a destination left ventricular device. J Heart Lung Transplant. 2010;29(12):1337–41.
59. Holman WL, Kormos RL, Naftel DC, Miller MA, Pagani FD, Blume E, et al. Predictors of death and transplant in patients with a mechanical circulatory support device: a multi-institutional study. J Heart Lung Transplant. 2009;28(1):44–50.
60. Cleveland JC, Naftel DC, Reece TB, Murray M, Antaki J, Pagani FD, Kirklin JK. Survival after biventricular assist device implantation: an analysis of the Interagency Registry for Mechanically Assisted Circulatory Support Database. J Heart Lung Transplant. 2011; 30(8):862–9.
61. Kusumoto FM, Goldschlager N. Device therapy for cardiac arrhythmias. JAMA. 2002;287(14):1848–52.
62. Goldberger Z, Lampert R. Implantable cardioverter-defibrillators: expanding indications and technologies. JAMA. 2006;295(7):809–18.
63. Young JB, Abraham WT, Smith AL, Leon AR, Lieberman R, Wilkoff B, et al. Combined cardiac resynchronization and implantable cardioversion defibrillation in advanced chronic heart failure: The MIRACLE ICD Trial. JAMA. 2003;289:2685–94.
64. Goldstein NE, Lapert R, Bradley E, Lynn J, Krumholz HM. Management of implantable cardioverter defibrillators in end-of-life care. Ann Intern Med. 2004;141(11):835–8.
65. Goldstein NE, Mehta D, Siddiqui S, Teitelbaum E, Zeidman J, Singson M, et al. "That's like an act of suicide": patients' attitudes toward deactivation of implantable defibrillators. J Gen Intern Med. 2007;23 Suppl 1:7–12.
66. Lampert R, Hayes DL, Annas GJ, Farley MA, Goldstein NE, Hamilton RM, et al. HRS expert consensus statement on the management of cardiovascular implantable electronic devices in patient nearing end of life or requesting withdrawal of therapy. Heart Rhythm. 2010;7(7):1008–26.
67. Goldstein N, Carlson M, Livote E, Kutner JS. Brief communication: management of implantable cardioverter-defibrillators in hospice: a nationwide survey. Ann Intern Med. 2010;152:296–9.
68. Gorodeski EZ, Chu EC, Reese JR, Shishehbor MH, Hsich E, Starling RC. Prognosis on chronic dobutamine or milrinone infusions for stage D heart failure. Circ Heart Fail. 2009;2:320–4.
69. MacIver J, Rao V, Delgado DH, Desai N, Ivanov J, Abbey S, et al. Choices: a study of preferences for end-of-life treatments in patients with advanced heart failure. J Heart Lung Transplant. 2008;27:1002–7.
70. Wachter RM, Luce JM, Hearst N, Lo B. Decisions about resuscitation: inequities among patients with different diseases but similar prognoses. Ann Intern Med. 1989;111:525–32.
71. Tanvetyanon T, Leighton JC. Life-sustaining treatments in patients who died of chronic congestive heart failure compared with metastatic cancer. Crit Care Med. 2003;31(1):60–4.
72. Chen JLT, Sosnov J, Lessard D, Goldberg RJ. Impact of do-not-resuscitate orders on quality of care performance measures in patients hospitalized with acute heart failure. Am Heart J. 2008;156(1):78–81.
73. Billings JA. Getting the DNR. J Palliat Med. 2012;15(12):1288–90.

74. Dev S, Clare RM, Felker GM, Fiuzat M, Stevenson LW, O'Connor CM. Link between decisions regarding resuscitation and preferences for quality over length of life with heart failure. Eur J Heart Fail. 2012;14:45–53.
75. Blinderman CD, Krakauer EL, Solomon MZ. Time to revise the approach to determining cardiopulmonary resuscitation status. JAMA. 2012;307:917–8.
76. Moons P, Bovijn L, Budts W, Belmans A, Gewillig M. Temporal trends in survival to adulthood among patients born with congenital heart disease from 1970 to 1992 in Belgium. Circulation. 2010;122(22):2264–72.
77. Bowater SE, Speakman JK, Thorne SA. End-of-life care in adults with congenital heart disease: now is the time to act. Curr Opin Support Palliat Care. 2013;7:8–13.
78. Tobler D, Greutmann M, Colman JM, Greutmann-Yantiri M, Librach LS, Kovacs AH. End-of-life in adults with congenital heart disease: a call for early communication. Int J Cardiol. 2012;155(3):383–7.
79. Goldfinger JW, Adler ED. End-of-life options for patients with advanced heart failure. Curr Heart Fail Rep. 2010;7:140–7.
80. Cheung WY, Schaefer K, May CW, Glynn RJ, Curtis LH, Stevenson LW, et al. Enrollment and events of hospice patients with heart failure vs. cancer. J Pain Symptom Manage. 2013;45:552–60.
81. National Hospice and Palliative Care Organization [Internet]. The Medicare hospice benefit. [cited 10 Aug 2013]. Available from: http://www.nhpco.org/sites/default/files/public/communications/Outreach/The_Medicare_Hospice_Benefit.pdf.
82. Wingate S, Bain KT, Goodlin SJ. Availability of data when heart failure patients are admitted to hospice. Congest Heart Fail. 2011;17:303–8.

Chapter 4
End-of-Life Care in Hospitalized Patients with Cardiovascular and Cerebrovascular Disease

Pablo Díez-Villanueva and Manuel Martínez-Sellés

Abstract We sought to review and increase understanding of the last days of life of patients hospitalized with advanced heart failure and acute stroke, which are the paradigms of end-of-life care of patients with cardiovascular and cerebrovascular disease, in order to improve treatment and clinical decision-making. These patients constitute an increasing health care problem, entailing both high morbidity and mortality. In-hospital death rates for cardiovascular illnesses have decreased over several decades, although they vary by geographic region and health care resources. Cardiovascular disease is the primary cause of death world-wide. While in-hospital death from coronary artery and valvular heart disease is less common now than it was several decades ago, reduction in hospital and short-term post-hospital mortality are major quality initiatives. Comprehensive clinical evaluation is essential in order to identify factors that may influence prognosis (e.g., comorbid conditions, functional status, and frailty) so that best and early care programs can be administered. Palliative care is known to be relatively undeveloped in these groups of patients, and should be implemented and be part of the multidisciplinary approach. The aim in patients likely to die is to improve symptoms and quality of life, while avoiding the use of aggressive therapies that are known to consume health-care resources without providing significant benefit. Physicians should not only use palliative measures when the patient is close to death, but throughout care. Dying patients with cardiovascular or cerebrovascular disease have the right to be involved in the decision-making process. Consequently, they or a designated surrogate should be fully informed about their medical condition, eventual interventions and prognosis. This is essential so that the can express their opinion and wishes, including place of death and psycho-spiritual care.

P. Díez-Villanueva, MD, PhD
Department of Cardiology, Hospital Universitario Gregorio Marañón,
Calle Doctor Esquerdo, 46, Madrid 28007, Spain

M. Martínez-Sellés, MD, PhD (✉)
Cardiology Department, Hospital Universitario Gregorio Marñón
and Universidad Europea, Madrid. Calle Doctor, Esquerdo, 46, 28007 Madrid, Spain
e-mail: mmselles@secardiologia.es

S.J. Goodlin, M.W. Rich (eds.), *End-of-Life Care in Cardiovascular Disease*,
DOI 10.1007/978-1-4471-6521-7_4

Keywords Advanced heart failure • Stroke survivors • End-of-life care • Hospitalized patients • Palliative care • Do-not-resuscitate orders • Implantable cardioverter-defibrillator deactivation

Key Points

- Patient selection and timeliness of procedures significantly impact hospital mortality for valvular and coronary heart disease.
- Heart failure patients variably die in the hospital, depending on geographic practice variations and available end of life care resources out of the hospital. Many heart failure patients are discharged and readmitted ultimately for a hospital death.
- Careful discussion of prognosis and attention to comorbidities, frailty and cognitive dysfunction can guide decisions about interventions in patients with advanced heart failure.
- End of life care for patients with stroke is important but few studies inform care.
- Depression has a great impact on the patient's symptoms for patients with heart failure and stroke, thus specific treatment is essential.
- Physicians ought to provide adequate control of patients' symptoms, especially fatigue, dyspnea, asthenia and pain.

Advanced heart failure (HF) and stroke are the paradigms of end-of-life care of patients with cardiovascular and cerebrovascular disease. HF is the final path in which converge most heart diseases, characterized by progressive disability ultimately leading to death. Stroke, on the other hand, has a different clinical trajectory, sharply changing patient's functional status, with likely dependencies. With increasing prevalence, HF entails an important healthcare problem, with high morbidity and mortality [1–4]. HF patients are often re-hospitalized and more than 60 % of patients admitted with HF have been previously diagnosed of chronic HF [5]. Patients with HF have a poor quality of life, particularly when they present symptoms that are resistant to treatment (stage D) [6]. The loss of functional capacity or autonomy may occur gradually, with a pattern of decompensation without complete recovery after each decompensation episode, or abruptly (i.e., in the setting of an acute myocardial infarction). It is often difficult to determine the real prognosis of the disease and anticipate the terminal phase, though life expectancy can be predicted in months in certain cases, such as in advanced stages.

In-hospital deaths from cardiovascular disease and stroke have decreased over the past few decades, as many deaths occur sometime after hospitalization. Interventions to treat coronary artery disease and valvular heart disease clearly have lower mortality rates in centers and surgery with more experience [7]. Coronary artery disease hospital mortality is impacted by time of patient presentation, with off-hours presentation associated with significantly higher mortality [8]. Mortality following

Table 4.1 Rates of in-hospital death by cardiovascular diagnosis

Diagnosis	Procedure	In-hospital mortality	
Coronary artery disease	Percutaneous coronary intervention	2.1–3.2 % [13]	
Coronary artery disease	Coronary artery bypass grafting	4.7–6.0 %	9.2 % mortality for age ≥ 80 years [14]
Heart failure		3.9 [15]–53 % [16]	
Aortic Valve Replacement	Transcutaneous	5 % at 30 days	
	Surgical	8 % at 30 days [17]	
Stroke		5.4 % (ischemic) [11]–27.2 % (intracranial hemorrhage) [18]	45.7 % had a poor outcome following discharge

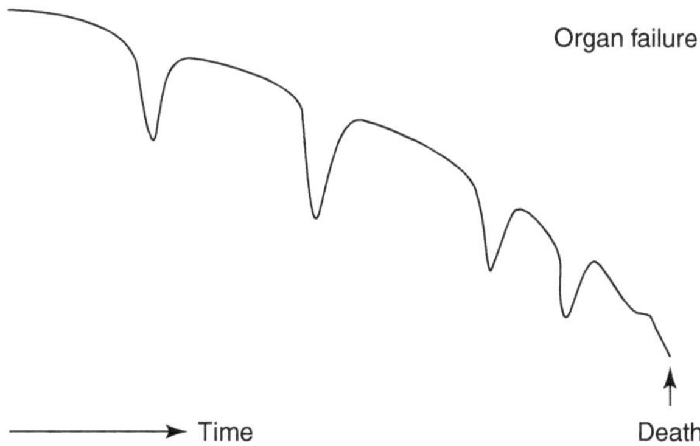

Fig. 4.1 Disease progression model in patients with organ failure

cardiovascular surgery is negatively impacted by increasing patient age [9]. Stroke in-hospital mortality is significantly higher for hemorrhagic than ischemic stroke [10] and rates for both etiologies are higher with increasing patient age, pre-stroke functional status, as well as stroke severity and complications [11].

Whether death occurs in the hospital may depend on regional health care resources. There is significant geographic variability in rates of in-hospital death and number of days of the terminal hospitalization for patients with cardiovascular disease and other chronic conditions [12]. Representative in-hospital mortality rates are presented in Table 4.1.

Dyspnea and fatigue are the cardinal manifestations of HF [19] but patients can also present improvement in symptoms, following the model of "organ failure" (Fig. 4.1) [20]. On the other hand, patients with HF have little knowledge of their disease or its treatment [21], and patients and their families frequently do not perceive it as being a serious disease [22]. All these factors can lead to optimism when estimating the risk and assessing life expectancy and may explain why patients in the final stages of HF frequently receive aggressive medical therapies, including intubation, resuscitation, and other measures only a few days before they die. Moreover HF is a disease of the elderly, with a mean age over 75 years and comorbidities may further worsen the prognosis [6, 19–23]. Although HF is known to be a life-shortening condition, equivalent to malignant disease in terms of symptom burden and mortality, it is not always easy to recognize that a patient is entering the final phase of the disease.

Due to the lack of studies referring to specific treatments for HF symptoms in advanced stages there is limited availability of data to guide physicians in the choice of treatments. In these advanced stages, goals like healing and prolonging life can be less important than reducing suffering and improving quality of life [24]. However, only a comparatively small number of HF patients receive specialist palliative care [23, 25, 26], which includes treatment of refractory symptoms as well as facilitates communication and decision making and family support.

Originally related to the care of patients with cancer, palliative care has expanded with the aim to improve the quality of life for patients and their families facing any life-threatening illness [27]. In patients with advanced HF, specific treatment and palliative care should be complementary. There is widespread recognition of the need for integration of palliative care in the care in these patients [19, 28–34]. Improvements in pharmacological, device, and cardiac surgical interventions have led a great number of patients with HF to live for many years following diagnosis, although mortality is still high. Advanced HF programs should involve cardiologists, general practitioners, HF nurses, and palliative care physicians.

Stroke, meanwhile, is one of the most important causes of death worldwide, as well as the leading cause of disability [35, 36]. The impact of stroke is variable and long lasting, and affects both the patient and their caregivers. However, only a few studies have focused on the symptomatic and palliative needs of these patients and their families [37], as most literature refers to acute management, secondary prevention and stroke rehabilitation [38–40]. End-of-life care is very important in acute stroke nursing due to high mortality rates in spite of advances in care [41], and patients with stroke have palliative care needs [42]. Multidisciplinary work should incorporate proper planning and care management as well as good communication with patient and family.

Which Patients Might Benefit from Palliative Care? Patient Selection

Although both health professionals and patients would like to define the prognosis of a subject with HF or acute stroke, the likelihood of survival can be reliably determined only in populations and not in individuals. However, estimating the prognosis of these patients may provide better information to patients and their families to help them to properly plan their future. Patients in stage D HF have a high mortality, with up to 75 % of deaths occurring in the year following the diagnosis and this information can guide decision-making and choice of treatment for these patients. Different analysis and studies performed during the last years have identified some parameters that can be used to assess the expectations of survival in patients with HF [43–45], summarized in Table 4.2. Many of these parameters have

Table 4.2 Key prognostic parameters in patients with heart failure

Key prognostic parameters
Advanced age
Left ventricular systolic dysfunction
Ischemic etiology
Functional class
Severe hyponatremia
Low peak oxygen consumption during exercise
Low hematocrit
Wide QRS complex
Hypotension
Sinus tachycardia at rest
Renal insufficiency
Intolerance to conventional therapy
Refractory volume overload
Persistently elevated natriuretic peptides
Comorbidity
Frailty

been shown to have independent predictive value on mortality in patients with HF and should be taken into account simultaneously, integrating them into predictive models. The CARING (Cancer, Admissions ≥2, Residence in a nursing home, Intensive care unit admit with multiorgan ≥2 Non-cancer hospice Guidelines) criteria have high sensitivity and specificity for death at 1 year [46], although this is a general index, and is not specifically designed for patients with HF. The criteria for approaching death most used are those of the National Hospice and Palliative Care Organization in the United States, which incorporates some specific criteria for patients with HF as well as some general guidelines that include likely death within 6 months, informed consent regarding symptom relief as a therapeutic aim, documentation concerning disease course, and under-nutrition. These criteria are applied as guidelines for Medicare-reimbursed hospice care; however they lack sensitivity or are inadequate when selecting patients, especially among the elderly [47]. Other criteria have also been proposed for admission to palliative care programs and models that have shown to predict mortality in patients with HF [36, 48–52]. However, they were not designed to select patients to palliative care, are complex, and were built from populations with few elderly patients, comorbid patients, women, and nonwhite races [53]. With the exception of the EFFECT tool [52], most of these models do not account for other co-morbidity, frailty or impairment of patient's functional status that have been proved to be independent prognostic factors [54–56], thus having significant limitations when predicting survival [30] particularly in the elderly [57]. Alternatively, patients who may benefit from a palliative approach may be identified using different criteria, based on easy-to-detect clinical characteristics from the time of admission onward, and thus facilitating the identification of people nearing the end of life regardless of the reasons for admission. The MAGGIC meta-analysis [58] identifies 13 independent predictors of mortality in HF, the most important factors being age, ejection fraction, serum creatinine, New York Heart Association (NYHA class) and diabetes. The score can be easily accessible at the website (www.heartfailurerisk.org). Finally, the patient's own perception of worsening health status has shown to predict a greater number of hospitalizations, an increase in mortality and greater consumption of resources irrespective of the presence of other poor prognostic factors [59].

Regarding cerebrovascular disease, larger strokes with major sequelae and disability, advanced age, previous dementia, previous or acquired depression and some other mental disorders, central post-stroke pain, fatigue, and comorbidities, frailty and worse functional status are the most important factors related to worse prognosis after stroke [37]. Those patients who survive to a stroke often have important physical disability in addition to significant psychological and social limitations [42], and would benefit from addition of palliative care.

Comorbidity, Frailty and Functionality as Prognostic Factors

Comorbidity is the rule in these patients. Its prevalence increases with age, contributing to a poor prognosis. The Charlson comorbidity index is an independent predictor of mortality for patients with HF [60]. Important aspects such as frailty and functional status, not included in most prognostic indexes, also impact the prognosis.

Frailty refers to the reduced ability to overcome physiological stress. Frailty is associated with a HF diagnosis and increased mortality [61]. Although frailty has many components, some criteria have been proposed to facilitate its diagnoses [30], summarized in Table 4.3. Functional status, the group of activities and functions needed to maintain autonomy in everyday physical, mental, and social function, is the single most important factor in predicting hospital mortality in the elderly person [62], surpassing other indexes of disease severity. Other measures of functional state, such as the Barthel basal index, have shown to be predictors of mortality in elderly patients hospitalized for HF [63]. Dependency in the instrumental activities of daily life, cognitive dysfunction, and symptoms of depression are independently associated with mortality in the elderly hospitalized due to medical disease [64]. As heart failure patients may prefer longevity over quality of life [65], it is important to incorporate palliation into heart failure care.

Table 4.3 Criteria proposed for frailty

Criteria
Weight loss (>10 % weight at 60 years or body mass index <18.5)
Lack of energy (≤3 on a scale ranging from 0 to 10 or feelings of being abnormally tired or weak during the previous month)
Limited physical activity (on an activity scale)
Reduced walking velocity (time spent in walking 4.6 m compared to age-adjusted speed)
Muscular weakness (as measured by strength test)

The presence of three or more of these signs or symptoms of frailty has been associated with a worse clinical course, with greater rates of dependency, hospitalization and death

Cognitive deterioration and depression are often present; 20–30 % of patients with HF [66] and one third of stroke survivors have clinically significant depression [67, 68] which worsens patients' perception of their state of health, impedes adequate rehabilitation and reduces both quality of life and survival. A recent study demonstrated that 9.1 % patients with an acute ischemic stroke had a history of dementia [69]. Such patients were older and had more severe strokes, and were also less likely to be admitted to a stroke unit or to receive thrombolysis. Those patients with preexisting dementia who did not die during a hospitalization for stroke had higher disability at discharge and were more frequently institutionalized after acute stroke. So, efforts to alleviate symptoms and improve functional status in such population are particularly important. Comorbidity also has an important role in the prognosis and management of patients with acute ischemic stroke [70].

Treatment and Prevention of Sudden Death: Do-Not-Resuscitate Orders

Management of sudden death is discussed in Chap. 7 (Dickinson). Electrical devices (see Lampert Chap. 11) are not appropriate for patients expected to die in the hospital, but symptomatic management of arrhythmias is sometimes appropriate. Amiodarone does not improve survival but as the only antiarrhythmic which has shown not to increase mortality in patients with heart disease [71], it may be an alternative to improve quality of life by reducing the number of arrhythmia episodes. Catheter ablation, an invasive procedure, could also be an alternative to a defibrillator in symptomatic patients to reduce the frequency of arrhythmic episodes [72]. Care for the patient with advanced heart disease and stroke should include a discussion of resuscitation (see Chap. 2- Tanner). Unfortunately, the do-not-resuscitate order, that taken strictly means not implementing cardiopulmonary resuscitation maneuvers, is often associated with a reduction in other treatment and care. After adjusting for the severity of the disease, prognostic factors and age, the patients with these orders are 30 times more likely to die than those without them [73], which may indicate a reduction in the quality of care. Patients and, sometimes, physicians are barely aware that the success rate of resuscitation after cardiopulmonary arrest is low (close to 20 %) [74, 75]. In the Study to Understand Prognosis and Preferences for Treatment (SUPPORT), the prevalence of do-not-resuscitate orders in patients admitted for HF was less than 5 % [75]. In fact, in a ward of patients with HF, physicians frequently have a mistaken view regarding the desire of the patient to receive cardiopulmonary resuscitation or not [51, 76, 77]. We cannot forget the fact that chronic HF patients often have thoughts about death, both during acute exacerbations and chronic stable phases, yet some of them do not feel comfortable considering or talking about their mortality.

The do-not-resuscitate decision should explicitly appear in the medical record of the patient after being agreed with the patient and, if possible, their family, and the medical team. We have previously studied the use of do-not-resuscitate orders and

Table 4.4 ICD deactivation

A. Steps suggested previous to deactivation of the ICD
Identification of cause and terminal stage of the situation
Identification of the risk and benefit of non-deactivation
Evaluate the capacity and competence of the patient to the decision
Choose the treatment option most consistent with the overall objective of the patient
Show respect to any decision of the patient
B. Optimal timings in which ICD-deactivation should be assessed
Previous to implantation
Clinical worsening
After repeated ICD firing
When stating do-not-resuscitate orders
In the final stage of life

palliative care in 198 consecutive deaths of patients with heart disease that occurred in the cardiology department of our hospital [78]. Interestingly, almost three-fifths of patients who died had a do-not-resuscitate order, although the decision to issue the order was frequently taken after administering aggressive treatment. This decision should be made after thorough assessment of the prognostic and quality of life indexes, and, in case of poor short-term prognosis, aggressive treatments could be avoided.

Regarding patients with acute stroke, families, and patients when possible, ought to be included in ongoing dialogue with physicians, ensuring that the death is both peaceful and dignified [79]. However, information regarding this topic is scarce. Also, it is well known that these patients have a poor prognosis after cardiopulmonary resuscitation maneuvers [80], which should also be considered. In any case, discussions between physicians and patients are especially important at this point [81]. However, there is great variability in the prevalence of do-not-resuscitate orders from one country to other, and even from some centers to others in the same country. A sensitive approach with good communication skills is of great importance in initiating discussion about prognosis of the advanced disease [34, 79].

Increasingly, physicians face patients in the terminal stage of an advanced disease with an implanted cardioverter defibrillator (ICD) device. In HF refractory to treatment or hospitalized with stroke and an ICD, it is appropriate to sensitively inform patients that the ICD treatment of arrhythmias may unnecessarily prolong the dying process and succumbing to a lethal arrhythmia may be a better mode of dying. Patients and their families should be reassured that ICD inactivation is not expected to immediately result in the death of the patient [34]. It is important to follow some steps previous to deactivation of the ICD [82], summarized in Table 4.4. Finally, the decision must be reported in the history (see also Lampert Chap. 11).

Palliative Care

Palliative care refers to those activities aimed at improving the quality of life of the patients and their families facing the problem of a potentially fatal disease, through the prevention and relief of suffering by identification, evaluation and treatment of pain and other physical, psychological, and spiritual problems [27]. This care includes supportive care, focused on alleviating symptoms, complications, and side-effects of interventions. Palliative care is not dependent on a specific health-care team [34], and should not be reserved for patients who are expected to die in a short period (Fig. 4.2), thus, it should be available to all patients needing comprehensive and integrated treatment along the whole disease trajectory, as summarized in Table 4.5. The lack of education among the professionals performing palliative care in HF, and vice versa, is worrying [83], particularly in a advances stages of the

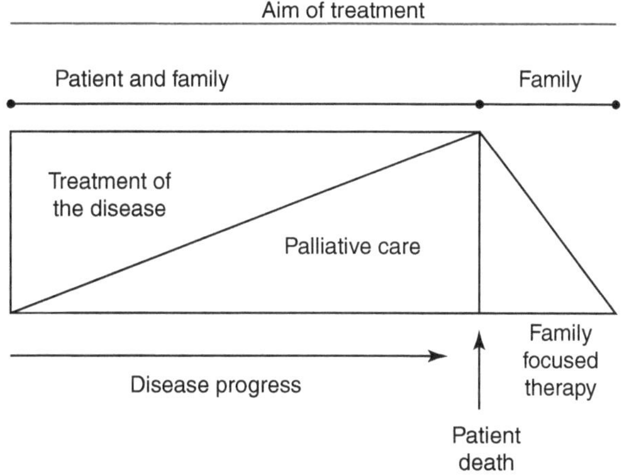

Fig. 4.2 Aim of treatment related to disease progression. Curative and palliative care can be combined in patients with advanced disease from the beginning

Table 4.5 Palliative care should include the following steps

Optimizing evidence-based therapy
Sensitively breaking bad news to the patient and family
Establishing an advanced care plan including documentation of the patients' preferences for treatment options
Education and counseling on relevant optimal self-management
Organizing multidisciplinary services
Identifying end-stage heart failure
Re-exploring goals of care
Optimizing symptom management at the end of life
Care after death including bereavement support

disease where "curative" or "life prolonging" therapeutic measures are increasingly ineffective [84, 85].

Following death, clinicians should acknowledge the grieving process among the family. Poor physician-family communication, being unprepared, failure to satisfy the family, poor symptom control and the lack of involvement of the health personnel during the last stage of life lead to pathological grieving.

As the aim of palliative care is to keep the patient as comfortable as possible, frequently treatment can be reviewed with the aim of simplifying it, since in patients close to death some drugs may be irrelevant. If the oral route of drug administration is not available, alternatives should be utilized, such as intravenous or subcutaneous routes [26]. In patients with advanced HF, death often occurs in hospital, even in those who have been treated at home for long periods [86, 87]. Some patients are concerned that dying at home would put too much stress upon their family, or family members may decline to consider death at home [88, 89]. One study conducted in UK demonstrated that no family member was offered the possibility of the patient dying at home with acute stroke [79]. It may be particularly important to provide caregiver training when patients likely to die with stroke are sent home [90–92].

Palliative Treatment of the Symptoms of Terminal Heart Disease

Scales to evaluate symptoms in patients with advanced HF, like the Memorial Symptom Assessment Scale (MSAS) [93], modified for HF (MSAS-HF) [94], and the Scale of the Edmonton Symptom Assessment (ESAS) [95]. In case of stroke patients, symptoms are often difficult to diagnose and consequently to treat, thus adversely impacting recovery, quality of life and mortality [37]. Patients with impaired communication need special attention [96]. Whatever scale is chosen it should be used to assess the symptoms throughout the course of the disease. A clinical interview should identify those factors that may worsen or improve symptoms, and, in the case of pain, its location and character. Palliative treatment involves diagnosing the cause or causes of each symptom to attempt to treat them. Symptoms need to be effectively evaluated as this is the primary target for treatment in the relief of suffering, but also because symptom characteristics may have prognostic implications in HF patients. Even if the cause of the symptom is irreversible, knowledge of the underlying mechanism should suggest the most appropriate symptomatic treatment [30].

Dyspnea

A total of 60 % of patients who die from advanced HF [97] and over 50 % of stroke survivor [37] suffer from severe dyspnea. Patients with advanced HF are known to have increased ventilation rates for a given volume of expired carbon dioxide thus causing tachypnea and increasing respiratory work, which is independent of symptomatic dyspnea. Pulmonary edema is associated with dyspnea. Together with optimal vasodilator and diuretic treatment, other likely causes should be investigated and treated, such as pleural effusion which can be alleviated by thoracocentesis. When dyspnea persists despite treatment, opioids should be used and these can lead to significant improvement of acute and chronic dyspnea in these patients, as well as ventilatory response [98]. Morphine sulfate or morphine chlorhydrate have the advantage of being able to be administered orally; doses range from 5 to 15 mg/4 h. (See also the detailed discussion in Palliative Care Chap. 3.) Oxygen therapy, even when the patient is not hypoxemic, as well as fresh air directed at the patient's face can also contribute to the relief of dyspnea. In stroke survivors, dyspnea is also related to older age, female sex and mental condition, and correlates with depression and disability [37], as well as with reduced long-term survival [99, 100]. Except for secondary causes like depression, obstructive sleep apnea, or hematologic, metabolic and endocrine disorders, there is not specific treatment [101].

Pain

Though its cause is not well defined, more than three out of four patients describe pain as the worst of their symptoms in the final phases of HF [102], and pain is severe 3 days before death in 41 % of patients. This pain may be due to cardiac causes, comorbidity (arthrosis, diabetic, or herpetic neuropathy, etc.) or to the medical treatment itself. Chest pain is common, also in the lower limbs and joints. Regardless of the cause, pain should be assessed and treated. Analgesics are the first line: paracetamol or low-dose opioids (codeine or dihydrocodeine).

About 25 % of stroke survivors, specially young and female patients, have chronic pain, which involves reduced quality of life [103]. Central post-stroke pain, hemiplegic shoulder pain, and painful spasticity and tension-type headache, are the most common types of pain [104]. The tricyclic antidepressant amitriptyline [105] is the drug of choice in central post-stroke pain, anticonvulsant agents [106] and gabapentin [107] are second-line drugs, whilst opioids are not effective [108]. Physical therapy is recommended in the others types of pain [37].

Sedation may be needed in some situations in order to control symptoms such as pain or dyspnea.

Depression

Depression should be distinguished from sadness or anticipatory grief, and from fatigue, insomnia and other symptoms that might be due to HF or stroke [37, 109]. Depression in hospitalized patients likely to die may respond to selective serotonin reuptake inhibitors however these drugs have been associated with hyponatremia or retention of fluids, probably due to the increase of antidiuretic hormone. On the other hand, tricyclic antidepressants have slow onset of action, a strong anticholinergic effect and can give rise to orthostatic hypotension, QTc prolongation, or delirium, so they should be avoided. There are some other effective measures for the treatment of depression, in addition to pharmacological treatment. Social and spiritual supports have demonstrated to be useful in reducing depression and anxiety [30, 110].

Asthenia

Cytokine and hormonal disturbances, characteristic of patients with advanced HF, have been associated with atrophy and weakness of the respiratory and skeletal muscles, thus being responsible for many symptoms such as fatigue and dyspnea as well as limitation of exercise capacity. Weakness may respond to more intense medical treatment and to an increase in physical activity if the patient is able to do this, as several studies have demonstrated the benefit of exercise on the quality of life of these patients.

Other Symptoms

The patients with HF may present some other symptoms like delirium, insomnia, nausea, vomiting, anorexia, weight loss, edema in the legs with ulcerations or cellulitis, constipation, diarrhea, anxiety, reduced mobility, or pruritus. Stroke patients may also present anxiety, emotionalism, incontinence, seizures and epilepsy [37]. These symptoms make the last stages of life more unpleasant and should be relieved using the most effective means.

The Role of Living Will. Dignity. Proportionate Use of Diagnostic and Therapeutic Measures in Terminal Cardiovascular and Cerebrovascular Diseases

Patients who know their diagnosis and prognosis and participate in decision-making have better symptomatic control. However the decision-making capacity of some patients with terminal heart disease may be impaired and their decision-making abilities should be evaluated. Patients are known to often have ambivalent and changing opinions [111, 112], with up to 40 % of patients with advanced HF changing their decision regarding the possibility of receiving resuscitation in less than 2 months [76], thus it is appropriate to readdress preferences on hospital admission.

Table 4.6 Chochinov's ABCD of care focused on conserving dignity

A. Attitude	A positive attitude within the health system has an impact on the attitude of the patient facing the disease
B. Behavior	To maintain dialogue, demonstrate interest, provide understandable daily information, take care during examinations, and respect modesty. Suggest having a family member present, especially if giving complex or "difficult" information, and address private subjects in an environment that respects privacy and from a nearby position
C. Compassion	The patient should perceive that the health personnel understand, respect, and empathize with their suffering
D. Dialogue	Dialogue should help us to know the patient as person. Knowing the vital state of the patient is basic to treatment focused on the conservation of dignity

Stroke patients rarely express a wish to die [37]. Living wills and other advance directives are useful in guiding decisions about care. Chochinov [113] suggests that the dying patient's perception of dignity depends to a great extent on their physician and proposes several approaches to care aimed at conserving dignity (Table 4.6).

Decision making, particularly about "futile" interventions is discussed in Chap. 5. Generally a futile intervention does not improve symptoms or intercurrent disease or potentially causes personal, familial, economic or social harm disproportionate to the expected benefit. The relationship between the risk involved in the measures to use and expected benefit has to be carefully analyzed. Clinician s should consider potential contribution to the patient's wellbeing, versus:

- worsening in the patient's quality of life,
- the risk of precipitating their death,
- pain or discomfort and
- prolongation of suffering by raising false hopes regarding improving survival
- in some settings the economic cost

Conclusions

Patients with terminal cardiovascular or cerebrovascular diseases have the right to care and attention in the last stages of their life. End-of-life care in such patients includes preventing needless suffering, weighing therapeutic risks and benefits, maintaining a trusting dialogue among the physicians, family members, and friends, and receiving symptomatic and psychospiritual support. Deciding that a patient is in the final phase of their disease requires correct evaluation of the characteristic prognostic factors of the disease itself and integration of comorbidities, frailty, functional, psychosocial, and nutritional status.

The aim of end of life care for hospitalized patients is to obtain the maximum quality of life for the patient based on a comprehensive approach. The organization of palliative care systems can improve the quality of life of patients before their death, and prevent the use of aggressive treatments that consume resources without having any effect on extending the patient's life, increasing patient satisfaction or that of the family.

References

1. Senni M, Tribouilloy CM, Rodeheffer RJ, Jacobsen SJ, Evans JM, Bailey KR, Redfield MM. Congestive heart failure in the community: trends in incidence and survival in a 10-year period. Arch Intern Med. 1999;159:29–34.
2. Hobbs FD, Kenkre JE, Roalfe AK, Davis RC, Hare R, Davies MK. Impact of heart failure and left ventricular systolic dysfunction on quality of life: a cross-sectional study comparing common chronic cardiac and medical disorders and a representative adult population. Eur Heart J. 2002;23:1867–76.
3. McMurray JJ, Adamopoulos S, Anker SD, Auricchio A, et al. ESC guidelines for the diagnosis and treatment of acute and chronic heart failure 2012: The Task Force for the Diagnosis and Treatment of Acute and Chronic Heart Failure 2012 of the European Society of Cardiology. Developed in collaboration with the Heart Failure Association (HFA) of the ESC. Eur J Heart Fail. 2012;14(8):803–69.
4. Yancy CW, Jessup M, Bozkurt B, Butler J, et al. 2013 ACCF/AHA guideline for the management of heart failure: executive summary: a report of the American College of Cardiology Foundation/American Heart Association Task Force on Practice Guidelines. J Am Coll Cardiol. 2013;62(16):1495–539.
5. Nieminen MS, Brutsaert D, Dickstein K, Drexler H, et al. EuroHeart Failure Survey II (EHFS II): a survey on hospitalized acute heart failure patients: description of population. Eur Heart J. 2006;27:2725–36.
6. Dickstein K, Cohen-Solal A, Filippatos G, McMurray JJ, Ponikowski P, Poole-Wilson PA, et al. Guias europeas de practica clinica para el diagnostico y tratamiento de la insuficiencia cardiaca aguda y cronica. Grupo de Trabajo de la ESC para el diagnostico y tratamiento de la insuficiencia cardiaca aguda y cronica (2008). Desarrollada en colaboracion con la Heart Failure Association (HFA) de la ESC y aprobada por la European Society of Intensive Care Medicine (ESICM). Rev Esp Cardiol. 2008;61:1329.e1–70.
7. Peterson ED, Coombs LP, DeLong ER, et al. Procedural volume as a marker of quality for CABG surgery. JAMA. 2004;291:195.
8. Sority A, Ahmed A, Starr SR, et al. Off-hours presentation and outcomes in patients with acute myocardial infarction: systemic review and meta-analysis. BMJ. 2014;348:1–16.
9. Martinez-Selles M, Hortal J, Barrio JM, Ruiz M, Bueno H. Treatment and outcomes of severe cardiac disease with surgical indication in very old patients. Int J Cardiol. 2007;119(1): 15–20.
10. Fonarow GC, Saver JL, Smith EE, Broderick JP, et al. Relationship of national institutes of health stroke scale to 30-day mortality in medicare beneficiaries with acute ischemic stroke. J Am Heart Assoc. 2012;1(1):42–50.
11. Koennecke HC, Beltz W, Berfelde D, et al. Factors influencing in-hospital mortality and morbidity in patients treated on a stroke unit. Neurology. 2011;77(10):965–72.
12. The Dartmouth Atlas of Health Care. Accessed at: http://www.dartmouthatlas.org/data/bar.aspx?ind=84&tf=10&ch=9,35&loc=&loct=3&fmt=109 on 04/01/14.
13. Cram P, Rosenthal GE, Vaughan-Sarrazin MS. Cardiac revascularization in specialty and general hospitals. N Engl J Med. 2005;352(14):1454.
14. Baskett R, Buth K, Ghali W, et al. Outcomes in octogenarians undergoing coronary artery bypass grafting. CMAJ. 2005;172:1183.

15. Hauptman PJ, Goodlin SJ, Lopatin M, Costanzo MR, Fonarow GC, Yancy CW. Characteristics of patients hospitalized with acute decompensated heart failure who are referred for hospice care. Arch Intern Med. 2007;167:1990–7.
16. Haga K, Murray S, Reid J, et al. Identifying community based chronic heart failure patients in the last year of life: a comparison of the Gold Standards Framework Prognostic Indicator Guide and the Seattle Heart Failure Model. Heart. 2012;98:579–83.
17. Smith CR, Leon MB, Mack MJ, Miller DC, PARTNER Trial Investigators, et al. Transcatheter versus surgical aortic-valve replacement in high-risk patients. N Engl J Med. 2011;364(23): 2187–98.
18. Smith EE, Shobha N, Dai D, Olson DM, Reeves MJ, Saver JL, Hernandez AF, Peterson ED, Fonarow GC, Schwamm LH. A risk score for in-hospital death in patients admitted with ischemic or hemorrhagic stroke. J Am Heart Assoc. 2013;2(1):e005207.
19. Goodlin SJ. Palliative care in congestive heart failure. J Am Coll Cardiol. 2009;54:386–96.
20. Lunney JR, Lynn J, Hogan C. Profiles of older medicare decedents. J Am Geriatr Soc. 2002;50:1108–12.
21. Martinez-Selles M, Garcia Robles JA, Munoz R, Serrano JA, Frades E, Dominguez Munoa M, et al. Pharmacological treatment in patients with heart failure: patients knowledge and occurrence of polypharmacy, alternative medicine and immunizations. Eur J Heart Fail. 2004;6:219–26.
22. Banegas JR, Rodriguez-Artalejo F. Insuficiencia cardiaca e instrumentos para medir la calidad de vida. Rev Esp Cardiol. 2008;61:233–5.
23. Goodlin SJ, Hauptman PJ, Arnold R, Grady K, et al. Consensus statement: palliative and supportive care in advanced heart failure. J Card Fail. 2004;10:200–9.
24. Morrison RS, Meier DE. Clinical practice. Palliative care. N Engl J Med. 2004;350: 2582–90.
25. Formiga F, López-Soto A, Navarro M, Riera-Mestre A, Bosch X, Pujol R. Hospital deaths of people aged 90 and over: end-of-life palliative care management. Gerontology. 2008;54: 148–52.
26. Formiga F, Chivite D, Ortega C, Casas S, Ramo'n JM, Pujol R. End-of-life preferences in elderly patients admitted for heart failure. QJM. 2004;97:803–8.
27. World Health Organization. WHO definition of palliative care. Geneva: World Health Organization; 2009. http://www.who.int/cancer/palliative/en/.
28. Hauptman PJ, Havranek EP. Integrating palliative care into heart failure care. Arch Intern Med. 2005;165:374–8.
29. Goodlin SJ. Palliative care for end-stage heart failure. Curr Heart Fail Rep. 2005;2:155–60.
30. Martínez-Selles M, Vidan MT, López-Palop R, Rexach L, Sánchez E, Datino T, et al. End-stage heart disease in the elderly. Rev Esp Cardiol. 2009;62:409–21.
31. Rexach L. Cuidados paliativos en insuficiencia cardiaca. Medicina Paliativa. 2010;17(5): 286–300.
32. Murray SA, Worth A, Boyd K, Kendall M, Hockley J, Pratt R, Denvir M, Arundel D. Patients', carers' and professionals' experiences of diagnosis, treatment and end-of-life care in heart failure: a prospective, qualitative interview study. London: Department of Health/British Heart Foundation Heart Failure Research Initiative. Final report; 2007.
33. Murray SA, Sheikh A. Palliative care beyond cancer: care for all at the end of life. Br Med J. 2008;336:958–9.
34. Jaarsma T, Beattie JM, Ryder M, Rutten FH, Advanced Heart Failure Study Group of the HFA of the ESC, et al. Palliative care in heart failure: a position statement from the palliative care workshop of the Heart Failure Association of the European Society of Cardiology. Eur J Heart Fail. 2009;11:433–43.
35. Jauch EC, Saver JL, Adams Jr HP, Bruno A, American Heart Association Stroke Council; Council on Cardiovascular Nursing; Council on Peripheral Vascular Disease; Council on Clinical Cardiology, et al. Guidelines for the early management of patients with acute ischemic stroke: a guideline for healthcare professionals from the American Heart Association/American

Stroke Association. Stroke. 2013;44(3):870–947. doi:10.1161/STR.0b013e318284056a. Epub 2013 Jan 31.
36. Morgenstern LB, Hemphill 3rd JC, Anderson C, et al. Guidelines for the management of spontaneous intracerebral hemorrhage: a guideline for healthcare professionals from the American Heart Association/American Stroke Association. Stroke. 2010;41:2108–29.
37. Creutzfeldt CJ, Holloway RG, Walker M. Symptomatic and palliative care for stroke survivors. J Gen Intern Med. 2012;27(7):853–60.
38. Langhorne P, Coupar F, Pollock A. Motor recovery after stroke: a systematic review. Lancet Neurol. 2009;8:741–54.
39. Langhorne P, Bernhardt J, Kwakkel G. Stroke rehabilitation. Lancet. 2011;377:1693–702.
40. Duncan PW, Zorowitz R, Bates B, et al. Management of adult stroke rehabilitation care: a clinical practice guideline. Stroke. 2005;36:e100–43.
41. Cowey E. End of life care for patients following acute stroke. Nurs Stand. 2012;26(27): 42–6.
42. Wee B, Adams A, Eva G. Palliative and end-of-life care for people with stroke. Curr Opin Support Palliat Care. 2010;4(4):229–32.
43. Aaronson KD, Schwartz JS, Chen TM, Wong KL, Goin JE, Mancini DM. Development and prospective validation of a clinical index to predict survival in ambulatory patients referred for cardiac transplant evaluation. Circulation. 1997;95:2660–7.
44. Pocock SJ, Wang D, Pfeffer MA, Yusuf S, McMurray JJ, Swedberg KB, et al. Predictors of mortality and morbidity in patients with chronic heart failure. Eur Heart J. 2006;27:65–75.
45. Felker GM, Hasselblad V, Hernandez AF, O'Connor CM. Biomarkerguided therapy in chronic heart failure: a meta-analysis of randomized controlled trials. Am Heart J. 2009;158: 422–30.
46. Fischer S, Gozansky WS, Sauaia A, Min SJ, Kutner JS, Kramer A. A practical tool to identify patients who may benefit from a palliative approach: the CARING Criteria. J Pain Symptom Manage. 2006;31:285–92.
47. Fox E, Landrum-McNiff K, Zhong Z, Dawson NV, Wu AW, Lynn J, SUPPORT Investigators. Evaluation of prognostic criteria for determining Hospice Eligibility in patients with advanced lung, heart or liver disease. JAMA. 1999;282:1638–45.
48. Casarett DJ, Quill TE. "I'm not ready for hospice": strategies for timely and effective hospice discussions. Ann Intern Med. 2007;146:443–9.
49. Huynh BC, Rovner A, Rich MW. Identification of older patients with heart failure who may be candidates for hospice care: development of a simple four-item risk score. J Am Geriatr Soc. 2008;56:1111–5.
50. Brophy JM, Dagenais GR, McSherry F, Williford W, Yusuf S. A multivariate model for predicting mortality in patients with heart failure and systolic dysfunction. Am J Med. 2004;116:300–4.
51. Levy WC, Mozaffarian D, Linker DT, Sutradhar SC, Anker SD, Cropp AB, et al. The Seattle Heart Failure Model: prediction of survival in heart failure. Circulation. 2006;113:1424–33.
52. Lee DS, Austin PC, Rouleau JL, Liu PP, Naimark D, Tu JV. Predicting mortality among patients hospitalized for heart failure: derivation and validation of a clinical model. JAMA. 2003;290:2581–7.
53. Heiat A, Gross CP, Krumholz HM. Representation of the elderly, women, and minorities in heart failure clinical trials. Arch Intern Med. 2002;162:1682–8.
54. Gijsen R, Hoeymans N, Schellevis FG, Ruwaard D, Satariano WA, van den Bos GA. Causes and consequences of comorbidity: a review. J Clin Epidemiol. 2001;54:661–74.
55. Afilalo J, Karunananthan S, Eisenberg MJ, Alexander KP, Bergman H. Role of frailty in patients with cardiovascular disease. Am J Cardiol. 2009;103:1616–21.
56. Lupon J, Gonzalez B, Santaeugenia S, Altimir S, Urrutia A, Mas D, et al. Prognostic implication of frailty and depressive symptoms in an outpatient population with heart failure. Rev Esp Cardiol. 2008;61:835–42.

57. Komajda M, Hanon O, Hochadel M, Lopez-Sendon JL, et al. Contemporary management of octogenarians hospitalized for heart failure in Europe: Euro Heart Failure Survey II. Eur Heart J. 2009;30:478–86.
58. Pocock SJ, Ariti CA, McMurray JJ, Maggioni A, Meta-Analysis Global Group in Chronic Heart Failure, et al. Predicting survival in heart failure: a risk score based on 39 372 patients from 30 studies. Eur Heart J. 2013;34(19):1404–13.
59. Heidenreich PA, Spertus JA, Jones PG, Weintraub WS, Rumsfeld JS, Rathore SS, et al. Health status identifies heart failure outpatients at risk for hospitalization or death. J Am Coll Cardiol. 2006;47:752–6.
60. Subramanian U, Eckert G, Yeung A, Tierney WM. A single health status question had important prognostic value among outpatients with chronic heart failure. J Clin Epidemiol. 2007;60:803–11.
61. Fried LP, Tangen CM, Walston J, Newman AB, Hirsch C, Gottdiener J, Cardiovascular Health Study Collaborative Research Group, et al. Frailty in older adults: evidence for a phenotype. J Gerontol A Biol Sci Med Sci. 2001;56:M146–56.
62. Lunney JR, Lynn J, Foley DJ, Lipson S, Guralnik JM. Patterns of functional decline at the end of life. JAMA. 2003;28:2387–92.
63. Formiga F, Chivite D, Casas S, Manito N, Pujol R. Valoracion funcional en pacientes ancianos ingresados por insuficiencia cardiaca. Rev Esp Cardiol. 2006;59:740–2.
64. Inouye S, Peduzzi PN, Robinson JT, Hughes JS, Howitz RI, Concato J. Importance of functional measures in predicting mortality among older hospitalised patients. JAMA. 1998;279:1187–93.
65. Brunner-La Rocca HP, Rickenbacher P, Muzzarelli S, Schindler R, TIME-CHF Investigators, et al. End-of-life preferences of elderly patients with chronic heart failure. Eur Heart J. 2012;33(6):752–9.
66. Rutledge T, Reis VA, Linke SE, Greenberg BH, Mills PJ. Depression in heart failure a meta-analytic review of prevalence, intervention effects, and associations with clinical outcomes. J Am Coll Cardiol. 2006;48:1527–37.
67. Chan W, Coutts SB, Hanly P. Sleep apnea in patients with transient ischemic attack and minor stroke: opportunity for risk reduction of recurrent stroke? Stroke. 2010;41:2973–5.
68. Hackett ML, Anderson CS, House A, Xia J. Interventions for treating depression after stroke. Cochrane Database Syst Rev. 2008;(4):CD003437.
69. Saposnik G, Cote R, Rochon PA, Mamdani M, Liu Y, Raptis S, Kapral MK, Black SE, Registry of the Canadian Stroke Network; Stroke Outcome Research Canada (SORCan) Working Group. Care and outcomes in patients with ischemic stroke with and without preexisting dementia. Neurology. 2011;77(18):1664–73.
70. Bernat JL. Do-not-resuscitate orders, quality of care, and outcomes in veterans with acute ischemic stroke. Neurology. 2013;81(2):199.
71. Bardy GH, Lee KL, Mark DB, Poole JE, Packer DL, Boineau R, Sudden Cardiac Death in Heart Failure Trial (SCD-HeFT) Investigators, et al. Amiodarone or an implantable cardioverter-defibrillator for congestive heart failure. N Engl J Med. 2005;352:225–37.
72. Reddy VY, Reynolds MR, Neuzil P, Richardson AW, Taborsky M, Jongnarangsin K, et al. Prophylactic catheter ablation for the prevention of defibrillator therapy. N Engl J Med. 2007;357:2657–65.
73. Ebrahim S. Do not resuscitate decisions: flogging dead horses or a dignified death? Resuscitation should not be withheld from elderly people without discussion. BMJ. 2000;320:1155–6.
74. Frank C, Heyland DK, Chen B, Farquhar D, Myers K, Iwaasa K. Determining resuscitation preferences of elderly inpatients: a review of the literature. CMAJ. 2003;169:795–9.
75. Goodlin SJ, Zhong Z, Lynn J, Teno JM, Fago JP, Desbiens N, et al. Factors associated with use of cardiopulmonary resuscitation in seriously ill hospitalized adults. JAMA. 1999;282:2333–9.
76. Krumholz HM, Phillips RS, Hamel MB, Teno JM, et al. Resuscitation preferences among patients with severe congestive heart failure: results from the SUPPORT project. Study to

Understand Prognoses and Preferences for Outcomes and Risks of Treatments. Circulation. 1998;98(7):648–55.
77. The SUPPORT Principal Investigators. A controlled trial to improve care for seriously ill hospitalized patients. JAMA. 1995;274:1591–8.
78. Martínez-Sellés M, Gallego L, Ruiz J, Fernández AF. Do-not-resuscitate orders and palliative care in patients who die in cardiology departments. What can be improved? Rev Esp Cardiol. 2010;63(2):233–7.
79. Payne S, Burton C, Addington-Hall J, Jones A. End-of-life issues in acute stroke care: a qualitative study of the experiences and preferences of patients and families. Palliat Med. 2010;24(2):146–53.
80. Bestué Cardiel M, Ara JR, Martín Martínez J. Do not resuscitate orders in patients with acute stroke. What circumstances decide their use? Med Clin (Barc). 2002;118(5):170–3.
81. Loertscher L, Reed DA, Bannon MP, Mueller PS. Cardiopulmonary resuscitation and do-not-resuscitate orders: a guide for clinicians. Am J Med. 2010;123(1):4–9.
82. Datino T, Rexach L, Vidán MT, Alonso A, Gándara A, Ruiz-García J, Fontecha B, Martínez-Sellés M. Guidelines on the management of implantable cardioverter defibrillators at the end of life. Rev Clin Esp. 2014;214:31–7.
83. Hauptman PJ. Palliation in heart failure: when less and more are more. Am J Hosp Palliat Care. 2006;23:150–2.
84. Gibbs JS, McCoy AS, Gibbs LM, Rogers AE, Addington-Hall JM. Living with and dying from heart failure: the role of palliative care. Heart. 2002;88 Suppl 2:ii36–9.
85. Abernethy AP, Currow DC, Frith P, Fazekas BS, McHugh A, Bui C. Randomised, double blind, placebo controlled crossover trial of sustained release morphine for the management of refractory dyspnoea. BMJ. 2003;327:523–8.
86. Formiga F, Olmedo C, Lopez SA, Pujol R. Dying in hospital of severe dementia: palliative decision-making analysis. Aging Clin Exp Res. 2004;16(5):420–1.
87. Roig E, Perez-Villa F, Cuppoletti A, Castillo M, Hernandez N, Morales M, et al. Programa de atencion especializada en la insuficiencia cardiaca terminal. Experiencia piloto de una unidad de insuficiencia cardiaca. Rev Esp Cardiol. 2006;59:109–16.
88. Lynn J, Teno JM, Phillips RS, Wu AW, Desbiens N, Harrold J, Claessens MT, Wenger N, Kreling B, Connors Jr AF. Perceptions by family members of the dying experience of older and seriously ill patients. SUPPORT Investigators. Study to Understand Prognoses and Preferences for Outcomes and Risks of Treatments. Ann Intern Med. 1997;126:97–106.
89. Willems DL, Hak A, Visser F, Van der Wal G. Thoughts of patients with advanced heart failure on dying. Palliat Med. 2004;18:564–72.
90. Han B, Haley WE. Family caregiving for patients with stroke. Review and analysis. Stroke. 1999;30:1478–85.
91. Van den Heuvel ET, de Witte LP, Schure LM, Sanderman R, Meyboomde Jong B. Risk factors for burn-out in caregivers of stroke patients, and possibilities for intervention. Clin Rehabil. 2001;15:669–77.
92. Kalra L, Evans A, Perez I, et al. Training carers of stroke patients: randomised controlled trial. BMJ. 2004;328:1099.
93. Blinderman CD, Homel P, Billings JA, Portenoy RK, Tennstedt SL. Symptom distress and quality of life in patients with advanced congestive heart failure. J Pain Symptom Manage. 2008;35:594–603.
94. Zambroski CH, Moser DK, Roser LP, Heo S, Chung ML. Patients with heart failure who die in hospice. Am Heart J. 2005;149:558–64.
95. Chang VT, Hwang SS, Feuerman M. Validation of the Edmonton Symptom Assessment Scale. Cancer. 2000;88:2164–71.
96. Kehayia E, Korner-Bitensky N, Singer F, et al. Differences in pain medication use in stroke patients with aphasia and without aphasia. Stroke. 1997;28:1867–70.
97. Levenson JW, McCarthy EP, Lynn J, Davis RB, Phillips RS. The last six months of life for patients with congestive heart failure. J Am Geriatr Soc. 2000;48:S101–9.

98. Williams SG, Wright DJ, Marshall P, Reese A, Tzeng BH, Coats AJ, et al. Safety and potential benefits of low dose diamorphine during exercise in patients with chronic heart failure. Heart. 2003;89:1085–6.
99. Appelros P. Prevalence and predictors of pain and fatigue after stroke: a population-based study. Int J Rehabil Res. 2006;29:329–33.
100. Mead GE, Graham C, Dorman P, Bruins SK, Lewis SC, Dennis MS, Sandercock PA, UK Collaborators of IST. Fatigue after stroke: baseline predictors and influence on survival Analysis of data from UK patients recruited in the International Stroke Trial. PLoS One. 2011;6:e16988.
101. McGeough E, Pollock A, Smith LN, et al. Interventions for post-stroke fatigue. Cochrane Database Syst Rev. 2009;(3):CD007030.
102. Pantilat SZ, Steimle AE. Palliative care for patients with heart failure. JAMA. 2004;291: 2476–82.
103. Jönsson AC, Lindgren I, Hallström B, Norrving B, Lindgren A. Prevalence and intensity of pain after stroke: a population based study focusing on patients' perspectives. J Neurol Neurosurg Psychiatry. 2006;77:590–5.
104. Klit H, Finnerup NB, Jensen TS. Central post-stroke pain: clinical characteristics, pathophysiology, and management. Lancet Neurol. 2009;8:857–68.
105. Leijon G, Boivie J. Central post-stroke pain—a controlled trial of amitriptyline and carbamazepine. Pain. 1989;36:27–36.
106. Vestergaard K, Andersen G, Gottrup H, Kristensen BT, Jensen TS. Lamotrigine for central poststroke pain: a randomized controlled trial. Neurology. 2001;56:184–90.
107. Serpell MG, Neuropathic Pain Study Group. Gabapentin in neuropathic pain syndromes: a randomised, double-blind, placebo-controlled trial. Pain. 2002;99:557–66.
108. Frese A, Husstedt IW, Ringelstein EB, Evers S. Pharmacologic treatment of central poststroke pain. Clin J Pain. 2006;22:252–60.
109. Pelle AJ, Gidron YY, Szabo BM, Denollet J. Psychological predictors of prognosis in chronic heart failure. J Card Fail. 2008;14:341–50.
110. Puchalski C, Ferrell B, Virani R, et al. Improving the quality of spiritual care as a dimension of palliative care: the report of the Consensus Conference. J Palliat Med. 2009;12:885–904.
111. Steinhauser KE, Christakis NA, Clipp EC, McNeilly M, McIntyre L, Tulsky JA. Factors considered important at the end of life by patients, family, physicians, and other care providers. JAMA. 2000;284:2476–82.
112. Drought TS, Koenig BA. "Choice" in end-of-life decision making: researching fact or fiction? Gerontologist. 2002;42:114–28.
113. Chochinov HM. Dignity and the essence of medicine: the A, B, C, and D of dignity conserving care. BMJ. 2007;335:184–7.

Chapter 5
End of Life Care in the Intensive Care Unit

Aluko A. Hope and Hannah I. Lipman

Abstract One in five Americans die in or directly following an ICU stay. Predicting who will survive the ICU is difficult and those who do survive struggle with post-ICU physical and psychological symptoms and functional and cognitive impairments. All patients, regardless of prognosis, benefit from a multidisciplinary, patient- and family-centered approach to care aimed at assessing and treating multiple sources of suffering. Research highlights potential areas for improvement in EOL care in the ICU. Measures of quality ICU palliative care and models for integrating palliative care into the ICU are reviewed. Skillful clinician communication is necessary to facilitate shared decision-making, provide care that is concordant with patients' values, and support patients facing serious life threatening illness and their families. Decision making and best practices for forgoing life sustaining treatment in the ICU are reviewed.

Keywords Palliative care • End-of-life care • Life sustaining treatment • Communication • Shared decision-making • Futility • Withdrawal of mechanical ventilator

Key Points

- ICU survivors struggle with physical and psychological symptoms and functional and cognitive impairments. Their families struggle with psychological symptoms.
- All ICU patients benefit from a multidisciplinary, patient- and family-centered approach to care aimed at assessing and treating multiple sources of suffering.

A.A. Hope, MD, MSCE
Division of Critical Care Medicine, Department of Medicine,
Montefiore Medical Center, Bronx, NY, USA

H.I. Lipman, MD, MS (✉)
Divisions of Geriatrics and Cardiology, Department of Medicine
and The Montefiore Einstein Center for Bioethics, Montefiore Medical Center,
111 E. 210th Street, Bronx, NY 10467, USA
e-mail: hlipman@montefiore.org

S.J. Goodlin, M.W. Rich (eds.), *End of Life Care in Cardiovascular Disease*,
DOI 10.1007/978-1-4471-6521-7_5

- Quality palliative care provided in the ICU includes: identifying a medical decision-maker, determining whether there is a relevant advanced directive, providing basic information about the ICU environment in the form of a pamphlet, offering an interdisciplinary family meeting within 5 days of admission, and offering social work and spiritual support.
- Two methods of increasing access to palliative care for ICU are: (1) Improving knowledge and skill of ICU staff under the direction of a clinician champion and/or the use of algorithms and order sets and (2) Using standard criteria to trigger proactive palliative subspecialty consultation.
- Critical care guidelines endorse a shared decision-making model in which the patient/family, experts on the patient's values, play an active role in decision-making with the clinical team, experts on prognosis and treatment options.
- Suggestions for skillful communication to facilitate shared decision making with patients and families are described.
- Treatment decisions, including decisions to forgo LST, should be made based on the benefits and burdens of the treatment in the context of the patient's prognosis and values.
- Although secular ethical principles hold no distinction between withholding and withdrawing LST, in practice, stopping a treatment already in place may feel different emotionally from not starting a treatment. When a LST is withdrawn, the intent is to remove an undesired treatment for which the burden outweighs the benefit or which is merely serving to prolong death from the underlying illness.
- The time limited trial is especially useful when prognosis is uncertain and allows the benefits of a particular treatment to be directly assessed, with the recognition that the treatment may be stopped if it does not achieve the patient's goals.
- Withdrawing LST requires the same careful preparation and quality expectations as all other treatments and procedures performed in the ICU.
- Guidelines suggest that once a decision has been made to withhold one LST (such as renal hemodialysis or vasopressors), clinicians should carefully consider the utility of other forms of LST.
- Interdisciplinary teams should aim for consensus about the decision to forgo LST.
- Differing perspectives about what are appropriate goals of treatment often underlie futility conflicts. Understanding the common reasons for futility conflicts is important for conflict resolution and consensus building.
- Suggested steps in withdrawing the ventilator, including treating symptoms before and after, addressed.

Introduction

As medical technologies improve and the population ages, more patients with chronic medical problems will be treated in an Intensive Care Unit (ICU). Although most ICU patients survive, one in five Americans die in or directly following an ICU stay [1] and many of these ICU deaths involve withdrawal or withholding of life-sustaining treatment (LST) [2]. Since the focus of the ICU is on sustaining life and restoring vital organ function, it can be a challenging environment in which to provide excellent end-of-life (EOL) care. Critically ill patients are often incapacitated and families are asked to be surrogate decision-makers in the face of their own grief and bereavement [3, 4]. Research highlights potential areas for improvement in EOL care in the ICU: many families have a very poor understanding of the diagnosis and prognosis of the ill patient [5] and continue to suffer high rates of depression, anxiety and post-traumatic symptoms [4, 6] in the months following the death of a loved one and often report clinician behavior contributing to their symptom burden [7, 8]. Families of patients who die want to be provided with frank information, want to feel that the appropriate amount of care was provided to the patient and they want both the patient and themselves to be treated respectfully and compassionately [9].

In this chapter, we describe models of EOL care in the ICU, synthesize relevant evidence regarding communication processes that best facilitate concordance between the treatments provided and the patients' values and preferences, and provide practical recommendations for forgoing LSTs in the ICU.

Integrating Palliative Care in the ICU

Predicting who will survive the ICU is difficult and ICU survivors often suffer with post-ICU physical and psychological symptoms [10–12], and functional and cognitive impairments [13]. Therefore, all patients, regardless of prognosis, benefit from a multidisciplinary, patient- and family-centered approach to care aimed at assessing and treating multiple sources of suffering. Such care should be integrated into the structure, process and culture of the ICU. Palliative care focuses on the assessment and treatment of complex symptoms, improving communication about care goals, promoting concordance between treatments provided and the patient's values and preferences and providing emotional support to the patient and the family [14–16]. Increasingly, ICUs are being called to integrate palliative care into the comprehensive care provided for all patients. Measures of quality ICU palliative care, based on qualitative research with patients and families [17] and expert opinion [18, 19] include identifying a medical decision-maker, determining whether there is a relevant advanced directive, providing basic information about the ICU environment in the form of a pamphlet, offering an interdisciplinary family meeting within 5 days of

Table 5.1 Domains of palliative care in the ICU and quality indicators

Domains of quality ICU palliative care [18][a]	Examples of ICU palliative care quality indicators [19]
Patient- and family-centered decision making	Identification of a medical decision maker (or other appropriate surrogate)
	Determine and document whether the patient has an advance directive
	Clarify the patient's resuscitation status
Communication with team and with patients/families	Providing families with an information leaflet
	Conduct an Interdisciplinary family meeting within 5 days of patient admission into the ICU
Symptom management and comfort care	Regular pain assessment during the ICU stay using an appropriate scales for communicative and non-communicative patients on a routine basis
	Manage pain optimally – standardize and follow best clinical practices regarding pain management in the ICU
Emotional and practical support for patients and families	Offering social work support to the patient/family
Spiritual support for patients and families	Offering spiritual support to the patient/family
Emotional and organizational support for ICU clinicians	Establish a staff support group based on input and needs of the ICU staff
Continuity of care	Prepare the patient and/or family for a change in clinician(s) and introduce new clinicians

[a]Identified by an interdisciplinary task force of experts in critical care and palliative care

admission, and offering social work and spiritual support (see Table 5.1) [20]. By integrating excellent palliative care into critical care, the ICU environment will be better able to support patients and families facing serious illness with high probability of death.

Depending on the institutional practice and culture and skill set of the ICU staff, palliative care may be provided by the members of the ICU clinical team (including the nurse, physician extenders like nurse practitioners or physicians' assistants, physicians, social workers, respiratory therapists) or by members of a palliative care consultation team which may include a nurse, physician, social worker, pastoral care provider and psychologist [21]. Palliative care consultative teams are present in a majority of acute care hospitals in the United States and the specialty is growing steadily in other countries [22].

There is a spectrum of different approaches for increasing access of ICU patients to palliative care. At either end of the spectrum, are two models, which are not mutually exclusive: a consultative model and an integrative model [21]. These are compared with the usual practice of palliative care consultation at the discretion of the treating team.

The consultative model aims to increase the involvement and effectiveness of specialty palliative care consultation with patients at the highest risk for poor outcomes [23–25]. Proactive palliative care consultation is triggered by the patient's baseline characteristics (for example, age $\geq$ 80 years or the presence of severe dementia), the patient's acute diagnoses (for example, global cerebral ischemia or prolonged multi-organ failure) or the patient's pattern of healthcare utilization prior to ICU admission (for example, hospital length of stay) [23–25]. This approach has been associated with decreased ICU length of stay, increased use of do not resuscitate (DNR) orders, and increased transitions to comfort care without any increase in mortality [23–25].

The integrative model aims to improve the care provided by the ICU team itself. ICU clinicians' palliative care knowledge and skills are improved under the direction of a clinician champion and/or by the use of systems (protocols or order sets) that have been shown to improve clinical practice [26]. For example, when ICU clinicians conduct routine, proactive meetings within 72 h of ICU admission with the families of high risk patients, ICU length of stay is reduced. This is primarily due to a shorter time to consensus regarding plan of care [27, 28].

Communication and Decision-Making

Skillful clinician communication is necessary to facilitate informed decision-making and to provide care that is concordant with the patient's values [29]. Effective communication goes beyond information disclosure and explores expectations, uses narrative to make sense of a complex situation, builds relationships by providing emotional support, explores role preferences and discusses concerns and conflicts openly (see Table 5.2) [29]. Because ICU patients are usually unable to participate in their own decision-making [3], communication usually occurs with families and surrogate decision makers [33].

Shared Decision-Making

Critical care guidelines endorse a shared decision-making model in which the patient, if able, and/or the family play an active role in decision-making [34, 35]. Shared decision making is based on a relationship of mutual respect between the clinical team, who are experts on prognosis and treatment options and the patient/family, who are experts on the patient's values. The goal is consensus regarding value-laden treatment decisions by discussing the nature of the decision to be made, exchanging relevant information and discussing preferred roles in decision-making [36]. The clinical team shares the burden of decision-making by being available to help clarify the patient's values and the impact of each option on the patient's goals.

Clinicians need guidance to operationalize shared decision making into practice [37]. A qualitative study of 51 family conferences in the ICU setting regarding EOL treatment decisions showed substantial variation in the frequency with which individual elements of shared-decision making occurred. Clinicians more frequently provided medical information and made efforts to understand the patient's treatment

Table 5.2 Essentials of effective communication in the ICU

Structure communication	Use available aids to ensure key elements of the encounter are accomplished: Mnemonic SPIKES provides a roadmap for giving bad news (Setting, Perception, Invitation, Knowledge, Empathy, Strategy and Summary) [30]
Engage the family	Use active listening skills to ensure families have adequate time to speak and ask questions "Ask-Tell-Ask" VALUE mnemonic also may increase engagement of family – **V**alue Comments made by the family; **A**cknowledge family emotions; **L**isten; **U**nderstand the patient as a person; **E**licit family questions [31]
Discuss prognosis	Offer to provide information about prognosis Acknowledge uncertainty in prognostic estimates Address prognosis for probable functional outcomes and survival Describe the recovery process and clinical milestones signaling improvement or deterioration
Attend to emotion	Use strategies to attend explicitly to emotions "NURSE" mnemonic [32] **N**ame **U**nderstand **R**espect **S**upport **E**xplore Reassure that comfort will be addressed and that patient/family will not be abandoned
Document the meeting	Summarize the discussion in the medical record
Debrief	Complex skill building requires reflection about what went well with the communication encounter and what could have been improved

preferences, but were less likely to ensure understanding of the medical facts or ask about role preferences in decision-making [37]. Multiple studies have now confirmed a wide variation of preferred roles in decision making among patients and families. Some prefer to retain total control, while others prefer to defer to clinicians [38–40]. (See Table 5.3 for specific suggestions on facilitating shared decision-making.)

Surrogate Decision-Making

Few patients prepare advanced directives specifying desired care in the event of incapacity (a living will) or authorize a specific surrogate decision maker (a health care proxy or durable power of attorney for health care) [44]. Even when present, these advance directives often do not anticipate the specific set of circumstances the patient faces in the ICU [45, 46] and may not diminish the substantial distress surrogates feel in having to make these decisions [45–47]. In a systematic review of 40 studies examining the effect on surrogates of making treatment decisions for

Table 5.3 Facilitating shared decision making in the ICU

Goal	Actions/behaviors	Examples
Understand the patient as a person	Explore: The patient's life story The patient's functional status prior to critical illness The patient's values and goals	"Tell me what life was like for him before he came into the hospital" "As he was getting sicker, what was most important to him"
Provide information about the patient's illness, treatment and prognosis	Elicit family understanding before providing information	"It's usually helpful to start by making sure we are all on the same page: What is your understanding of the medical situation at this point?"
	Avoid medical jargon	
	Don't talk too much and allow time for patients and families to speak	
Provide information about patient's prognosis	Ask permission to provide information about prognosis	"Would you like me to talk about what we think is going to happen?"
	Be transparent about uncertainty	"I wish I had better news for you"
	Consider using an "I wish" statement [41]	
Check for understanding and interpretation of medical information	Explore the family's interpretation of the medical facts	"What questions do you have about what I have just said" "Does what I have just said fit with what you are seeing as well?"
Help surrogates understand their role	Normalize the surrogates' role in complex EOL decisions by emphasizing how common such decisions are in the modern ICU [42]	"Many families are called upon the help make decisions for their loved ones"
	Explain that decisions are value laden	"Different people feel very differently about what kind of treatments they would accept if they became very sick…."
	Explain substituted judgment	"Our goal is to honor your mother by trying to understand what she would choose if she were able" "Sometimes it is really hard to separate what you want from what your mother would want"
Explore specific values and value conflicts	Be able to name specific values that can be in conflict in EOL decision-making (for example, living as long as possible, avoiding physical/cognitive decline, maintaining independence [43]	"It sounds like it was important to him to live as independently as possible for as long as possible, is that right"
Discuss preferred decision-making styles	Understand that some families will want more guidance than others	"I want to give you as much guidance as you think you need regarding this decision. Some families prefer to take the information and make the decision themselves; other families prefer me to provide a recommendation"
	Be prepared to provide a recommendation based on the patients' values	

(continued)

Table 5.3 (continued)

Goal	Actions/behaviors	Examples
Attend to spirituality	Explore the spiritual needs of the patient	"It sounds like his faith is very important to him"
	Explore the role of spiritual beliefs in surrogate decision-making near EOL	"How does your faith impact how you see things?"
		"I can see that your faith is a tremendous help to you"
		"What would a miracle look like for him?"
Elicit concerns and questions		"You just got a lot of information. What questions do you have?
		"What concerns do you have? What is unclear?"

incapacitated adults, 27 of which focused on EOL treatment decisions, Wendler et al. reported that surrogates frequently experience significant stress, anxiety, guilt and moral distress [47]. In one prospective cohort study of surrogate decision-makers of incapacitated adults in ICUs, researchers found that 27 % of the sample had relatively low confidence in their ability to act as a surrogate [48]. Lower surrogates' confidence in their role was associated with poorer quality of communication between the clinician and the family, suggesting that better communication may improve surrogates' confidence. Surrogates may misconstrue their role. They may underestimate the importance of their input regarding the patient's values and beliefs or believe the decision should be based on their own values and desires rather than those of the patient [49]. Clinicians often miss opportunities to explain surrogate decision making [50]. Clinicians should clarify sensitively that the patient's values should drive EOL treatment decisions and acknowledge the difficulty of being a surrogate decision-maker, especially when patient preferences conflict with the family's emotional needs.

Information Disclosure

In order to ensure the patient and family understand the patients' diagnoses and prognosis, clinicians must convey information clearly and avoid medical jargon. Information may need to be repeated over time as integrating complex information under stressful circumstances is difficult. Delivery of information should be structured in order to maximize opportunities for family members to speak and ask questions [51]. An increased proportion of family speech during family meetings is significantly associated with increased satisfaction with clinician communication [51]. One recommended strategy for information disclosure is "Ask-Tell-Ask." The clinician begins the discussion by asking patients or surrogates to describe their understanding, including what they have been told by other clinicians, and asking permission to provide them information. The clinician then continues with a concise update in layperson's terms. The family is then asked to summarize what was discussed and are given the opportunity to ask questions or address their concerns [52].

Clinicians may be reluctant to provide information about prognosis or expected outcomes out of concern they would be judged negatively for the inherent uncertainty of the information or that they may extinguish hope or worsen emotional distress [53]. However, patients' families and surrogates generally desire frank information about prognosis [54], recognize the uncertainty of prognostic estimates and are interested in using prognostic estimates to emotionally and logistically prepare for potential worst outcomes [55, 56]. The process of surrogates preparing and accepting the possibility of death as an outcome is a slow and incremental one [54–56], rife with intrapersonal conflicts and tensions [57, 58] and often limited by psychological biases that make interpreting grim prognosis more difficult for patients and surrogates [59]. In a mixed qualitative and quantitative study in which 80 surrogates of critically ill patients were asked to use a probability scale to interpret sixteen prognostic estimates, researchers found that surrogates were able to accurately interpret statements expressing a high probability of a good outcome but were statistically more optimistic in their interpretation of statements that conveyed a high risk of death [59]. Techniques to enhance surrogate understanding of grim prognostic estimates include: checking that surrogates are ready to hear the information, giving adequate forewarning that the news is not good, acknowledging uncertainty and devoting time to exploring surrogates' perceptions [60]. In one qualitative study of 179 surrogates' perception of their critically ill family member's chance of survival, most appeared to weigh the clinician's judgment against their own knowledge of the patient's strength of character, their own observations of the patient while in the ICU, and their own optimism, intuition and religious faith [55]. By sensitively listening to surrogates explain their world view, clinicians can build trust and may decrease the potential for conflict.

Attending to Affect and Emotion

Research in decision psychology suggests that strong emotions and affect impact reasoning and information processing [61]. Specifically, emotions can change people's estimates of both value and risk. For example, when emotions are heightened, people are more likely to substitute how they feel about a particular outcome for a more logical reasoned assessment [61]. In addition, people appear to have difficulty forecasting what they will want and how they will behave in affective states different from their current state (so called, "hot-cold empathy gaps") [62]. This has relevance to EOL care in the ICU in several ways: surrogates are often making decisions under stress and while grieving over bad news; patients previously expressed wishes were often made when they were in relatively cool affective states making it hard for surrogates to be sure of what they would have wanted during the stress of illness. Despite the extreme importance of emotions and affect in decision-making, clinicians often miss opportunities to acknowledge and address emotions [50, 63]. Strategies for attending explicitly to emotions are listed in Table 5.2.

Attending to Spirituality

Patients' and families' religious and spiritual values may impact how they make sense of illness and approach EOL decisions. Tolerance and respect for religious and spiritual beliefs should be conveyed [64]. Surrogates with strong spiritual beliefs express more doubt about the accuracy of physician prognostication [65]. When surrogates doubt the clinician's prediction of poor prognosis, they are more likely to request continued life support [66]. Families may cite spiritual or religious explanations for why they are not willing or ready to discuss transitioning care from restorative to comfort-oriented. In general, such explanations should be seen as an opening for the sensitive exploration of the patient's or family's spiritual needs and beliefs. Tools are available for assessing spirituality in the patient encounter [67–69] and reviews have been written for approaching a patient or family with expectations of a miracle [70], or responding to a request for prayer or other spiritual ceremonies [71].

Communication Aids

Clinician communication can be supplemented with printed, online or video informational material which may be helpful in improving family members' understanding of complex medical information [31, 72, 73]. In a multicenter randomized clinical trial to improve the effectiveness of information provided to family surrogates in ICU, standard information (at least one meeting with the physician during the first week of ICU admission) was compared with standard information plus an information leaflet [72]. Family members in the standard communication arm expressed poor comprehension of the diagnosis, prognosis or treatment (as measured by recall questionnaires very soon after the family meeting) about 40 % of the time. This was decreased to 11 % with the addition of the information leaflet [72]. Brochures have been developed to help families prepare for face to face meetings with the clinical team [74] and better understand the syndrome of chronic critical illness, the hallmark of which is the prolonged dependence on mechanical ventilation and other intensive care therapies [75, 76]. Video decision support tools are being studied as an adjunct to advance care planning discussions. A small randomized clinical trial showed that patients with advanced cancer, who viewed the video in addition to communicating verbally with the clinician, reported improved knowledge about CPR and an increased preference for comfort-oriented care near EOL [77]. In a feasibility pre-post study, a CPR video support tool for surrogates in the ICU setting was associated with improved CPR knowledge without any change in their CPR preferences [78]. More research is needed on how to integrate technologies like video and internet with direct clinician communication in order to improve information understanding, relationship building and ultimately improve patient and surrogate outcomes.

Forgoing Life Sustaining Treatment: Principles and Practice

Ethical Principles

Patients with decision making capacity may choose to forgo (withdraw or withhold) any treatment. There is no ethical distinction made among various LSTs, such as mechanical ventilation, extracorporeal membrane oxygenation, left-ventricular assist devices, vasopressors, dialysis, antibiotics, blood products, intravenous fluids or artificial nutrition. The decision is made based on the benefits and burdens of the treatment in the context of the patient's prognosis and values. The patient's values may dictate that LST would be undesirable if it merely served to prolong life in a state s/he considers a fate worse than death or the benefit does not justify the burden of treatment. Clinicians can help patients and families navigate decisions about LST by clarifying prognosis and the expected benefit of treatment and exploring patient's values. Decisions by surrogates to forgo LST on behalf of incapacitated patients are ethically justifiable when made according to patient's wishes, if known, or in patient's best interests. The legal authority and responsibility of surrogate decision makers regarding forgoing LST vary from state to state in the United States. Clinicians must be familiar with applicable local laws and regulations.

Although secular ethical principles hold no distinction between withholding and withdrawing LST [79, 80], in practice, stopping a treatment already in place may feel different emotionally from not starting a treatment. Withdrawing a treatment may be emotionally more difficult, especially when death follows soon after, even though the intent is to remove an undesired treatment for which the burden outweighs the benefit or which is merely serving to prolong death from the underlying illness. Particularly when there is significant uncertainty about prognosis, not starting a treatment may be more difficult and leave unresolved the question of whether the patient could have benefited. Reluctance to start a particular treatment in order to avoid deciding to discontinue it later deprives families and clinicians of a time limited trial of therapy [15]. The time limited trial is especially useful when prognosis is uncertain and allows families and clinicians to directly assess the benefits of a particular treatment with the recognition that the treatment may be stopped if it does not achieve the patient's goals [15].

Clinical Principles

Withdrawing LST requires the same careful preparation and quality expectations as all other treatments and procedures performed in the ICU. The communication process and the specific plan for the withdrawal of LST should be documented in the chart. Novices should be supervised by expert clinicians, given direct feedback and opportunities for debriefing. Clinicians in training should not be allowed to withdraw LST independently until they have demonstrated competence in a supervised setting.

The goal of withdrawing LST is to remove treatments that are no longer wanted or do not provide any comfort to the patient. In practice, clinicians tend to withdraw treatments in a non-random sequence (for example, withdrawal of mechanical ventilation tends to occur after forgoing dialysis or vasopressors) [81, 82]. This sequential approach (also called stuttering withdrawal) may help clinicians and families create psychological distance between their decisions and the inevitable death and may make it easier for both clinicians and families to place the responsibility for the death on the underlying illness or failing body [83]. In general, guidelines suggest that once a decision has been made to withhold one LST (such as renal hemodialysis or vasopressors), clinicians should carefully consider the utility of other forms of LST.

Interdisciplinary teams should aim for consensus about the decision to forgo LST. All team members with direct patient care roles should feel included in decision making. Clinicians' decisions to limit LSTs are based on patient characteristics and influenced by multiple non-clinical, idiosyncratic factors [84–86]. A survey of clinicians across multiple university hospitals in Canada showed, that clinicians were more likely to withdraw LST from older patients, patients who they felt were unlikely to survive the current illness episode, and those who had poor baseline cognitive and physical function [85]. Provider level factors like the clinician's age, number of years since completing training, sub-specialty training, gender, race, religious beliefs may underlie variability in clinician's attitudes about withdrawal of life sustaining therapies [84, 87, 88]. These variations in approach to forgoing LST may lead to conflict among team members about decisions for individual patients. An approach which emphasizes consensus building among the interdisciplinary team is important for conflict prevention and management [89].

Futility

The concept of futility is often invoked when the clinician determines an intervention is inappropriate and the patient or family disagrees. Futility should be defined in terms of the goals of the treatment. As the term is used variably, the question of what goal the treatment is futile to achieve must be answered. The notion of *quantitative* futility deals strictly with the question of physiologic impact. A quantitatively futile treatment is one that has a very small probability of achieving its physiological purpose. A commonly proposed threshold is 1 %, i.e. a treatment that fails 99 times out of 100 is quantitatively, or physiologically, futile [90]. Rarely is a futility conflict only about the likelihood the treatment will achieve its physiologic purpose. *Qualitative* futility deals with the question of whether a treatment achieves acceptable goals. Differing perspectives about what are appropriate goals of treatment often underlie futility conflicts. Declaring a treatment futile is often how the clinician attempts to wrestle control over decision making from patients and families. Understanding the common reasons for futility disputes may make possible an alternate approach – conflict resolution and consensus building. Common reasons for conflict include:

- misunderstanding of the medical facts, including prognosis and the nature and goals of treatment options
- surrogates' emotional stressors, such as grief, guilt, mistrust, avoiding responsibility for end of life decision making or fear of abandonment
- conflicts of interest such as secondary gain (patient/family) or resource and/or time pressures (clinical team)
- intrateam or intrafamily conflict
- fundamental differences in values about what is life worth living and/or what are appropriate goals of medicine [91, 92].

Futility conflicts are best addressed by a process-oriented approach, informed by the common underlying causes of such conflicts, that utilizes best communication practices to clarify misunderstandings, support surrogates in emotional distress, recognize and minimize conflicts of interest, elucidate values and negotiate achievable goals of care in order to achieve consensus [93].

Practical Guidance

The ICU room should be transformed, as much as possible, from one equipped for rescue to one more conducive to comfort, dignity and quiet. Turning off unneeded ICU monitors allows the family to attend to the patient rather than the vital signs. Tubes, lines and drains that can be removed without patient discomfort should be removed, curtains should be closed and visitation hours liberalized to allow family the chance to say goodbye.

Most families will not know what to expect during this process. Provide key information to help alleviate their fears, worries and concerns. Acknowledge that the prognosis and time to death after the withdrawal of LST varies according to the clinical situation and may not be predicted accurately [94, 95]. Estimates based on the clinical condition of the patient (for example, minutes to hours or days to weeks) are helpful for help families who request more specific guidance. Families often will have reasons to delay, for example waiting for a family member from out of town or wanting to prepare funeral arrangements ahead of the death. If death is not expected to be imminent after the withdrawal procedure, palliative care consultation teams may be helpful in facilitating transition to a private room in the hospital or to a hospice setting. The possibility of transfer out of the ICU should be discussed with families openly before it is initiated. For families who would like to stay at the bedside during the withdrawal procedure, the clinician should feel comfortable describing the procedures, particularly the possibility of noisy respirations, changes in breathing patterns, possibility of delirium, muscle twitching or other idiosyncratic movements. Families should be offered spiritual and emotional support by a chaplain or social worker respectively.

Because the withdrawal of mechanical ventilation has the greatest potential for patient discomfort, we offer practical suggestions below. Usually all other LST are withdrawn prior to removing the mechanical ventilator. After adequate sedation has been achieved, LST like vasopressors, pacemakers, intraortic balloon pumps can simply be turned off without any tapering.

Withdrawing the Mechanical Ventilator

There is no optimal strategy for removing the mechanical ventilator from the patient and practice varies. Some clinicians prefer to withdraw the mechanical ventilator and the endotracheal tube in one step. Others prefer to maintain the artificial airway to facilitate suctioning and prevent obstruction. Limited data suggest that critical care survivors found suctioning via endotracheal tube very uncomfortable [96]. Therefore, most patients should have the endotracheal tube removed with adequate analgesia and sedation.

Effective sedation and analgesia is crucial. Ensuring the comfort of the patient is the primary goal. Both the "preemptive" phase and the "reactive" phase are important for effective symptom control [97]. For example, a patient with acute respiratory distress syndrome on maximum ventilator support may experience worsening symptoms when positive end expiratory pressure or the oxygen concentration is decreased. Additional sedation and analgesia may need prior to the initiation of the withdrawal process and again after the ventilator is removed in response to the patient's symptoms.

Suggested steps in withdrawing the mechanical ventilator:

1. Pre-emptive phase of analgesia and sedation plan in order to ensure the patient is comfortable before initiating the withdrawal of the ventilator
2. Assess the likelihood of imminent death after the withdrawal of the ventilator
3. Have available at the bedside medications for sedation and analgesia
4. Consider turning off monitor alarms
5. Reduce ventilator support (titrate down FiO_2, pressure support and PEEP) over a relatively short time period (e.g. 5–30 min).
6. Reassess and treat any symptoms
7. Extubate, oral suction as needed, monitor closely
8. Reassess treat any symptoms, especially pain, dyspnea, delirium, oropharyngeal secretions

The use of pre-emptive general anesthesia (high doses of opioids and sedatives to accomplish deep sedation) is not standard of care but might be appropriate when the risk of symptom distress in a conscious patient is high and this approach to symptom management is consistent with the patient's values and the clinicians' skill [97, 98]. Neuromuscular blocking agents should not be started during the process of withdrawing the mechanical ventilator. Reports suggest that a minority of physicians continue to use these drugs to facilitate a comfortable "look" during the withdrawal process [99, 100]. Since these drugs do not have any analgesic or sedative properties, their use cannot be justified as part of a comfort oriented approach. Furthermore, paralysis prevents the patient from displaying symptoms, but not from experiencing them. In most instances, if a paralytic agent was used for therapeutic purposes prior to the withdrawal of the mechanical ventilator, a reasonable amount of time should be provided to allow the restoration of neuromuscular function to facilitate symptom assessment during the withdrawal of mechanical ventilation in these patients. In rare instances, when restoring neuromuscular function is not feasible, withdrawal may proceed with adequate doses of pre-emptive analgesia and sedation to ensure the comfort of the patient during the dying process.

Depending on the patient's prognosis, intense nursing support may be required to achieve symptom control after extubation. Since most of the patients will not be able self-report symptoms, clinicians should be familiar with tools that have been developed for the behavioral assessment of pain and dyspnea [101, 102]. A full review of the management of EOL symptoms is beyond the scope of this chapter but has been reviewed elsewhere [100, 103–107].

References

1. Angus DC, Barnato AE, Linde-Zwirble WT, Weissfeld LA, Watson RS, Rickert T, et al. Use of intensive care at the end of life in the United States: an epidemiologic study. Crit Care Med. 2004;32(3):638–43.
2. Prendergast TJ, Luce JM. Increasing incidence of withholding and withdrawal of life support from the critically ill. Am J Respir Crit Care Med. 1997;155(1):15–20.
3. Prendergast TJ, Claessens MT, Luce JM. A national survey of end-of-life care for critically ill patients. Am J Respir Crit Care Med. 1998;158(4):1163–7.
4. Pochard F, Azoulay E, Chevret S, Lemaire F, Hubert P, Canoui P, et al. Symptoms of anxiety and depression in family members of intensive care unit patients: ethical hypothesis regarding decision-making capacity. Crit Care Med. 2001;29(10):1893–7.
5. Azoulay E, Chevret S, Leleu G, Pochard F, Barboteu M, Adrie C, et al. Half the families of intensive care unit patients experience inadequate communication with physicians. Crit Care Med. 2000;28(8):3044–9.
6. Azoulay E, Pochard F, Kentish-Barnes N, Chevret S, Aboab J, Adrie C, et al. Risk of post-traumatic stress symptoms in family members of intensive care unit patients. Am J Respir Crit Care Med. 2005;171(9):987–94.
7. Tilden VP, Tolle SW, Garland MJ, Nelson CA. Decisions about life-sustaining treatment. Impact of physicians' behaviors on the family. Arch Intern Med. 1995;155(6):633–8.
8. Abbott KH, Sago JG, Breen CM, Abernethy AP, Tulsky JA. Families looking back: one year after discussion of withdrawal or withholding of life-sustaining support. Crit Care Med. 2001;29(1):197–201.
9. Heyland DK, Rocker GM, O'Callaghan CJ, Dodek PM, Cook DJ. Dying in the ICU: perspectives of family members. Chest. 2003;124(1):392–7.

10. Nelson JE, Meier DE, Litke A, Natale DA, Siegel RE, Morrison RS. The symptom burden of chronic critical illness. Crit Care Med. 2004;32(7):1527–34.
11. Puntillo KA, Arai S, Cohen NH, Gropper MA, Neuhaus J, Paul SM, et al. Symptoms experienced by intensive care unit patients at high risk of dying. Crit Care Med. 2010;38(11):2155–60.
12. Myhren H, Ekeberg O, Toien K, Karlsson S, Stokland O. Posttraumatic stress, anxiety and depression symptoms in patients during the first year post intensive care unit discharge. Crit Care. 2010;14(1):R14.
13. Desai SV, Law TJ, Needham DM. Long-term complications of critical care. Crit Care Med. 2011;39(2):371–9.
14. Lanken PN, Terry PB, Delisser HM, Fahy BF, Hansen-Flaschen J, Heffner JE, et al. An official American Thoracic Society clinical policy statement: palliative care for patients with respiratory diseases and critical illnesses. Am J Respir Crit Care Med. 2008;177(8):912–27.
15. Truog RD, Campbell ML, Curtis JR, Haas CE, Luce JM, Rubenfeld GD, et al. Recommendations for end-of-life care in the intensive care unit: a consensus statement by the American College of Critical Care Medicine. Crit Care Med. 2008;36(3):953–63.
16. Selecky PA, Eliasson CA, Hall RI, Schneider RF, Varkey B, McCaffree DR. Palliative and end-of-life care for patients with cardiopulmonary diseases: American College of Chest Physicians position statement. Chest. 2005;128(5):3599–610.
17. Nelson JE, Puntillo KA, Pronovost PJ, Walker AS, McAdam JL, Ilaoa D, et al. In their own words: patients and families define high-quality palliative care in the intensive care unit. Crit Care Med. 2010;38(3):808–18.
18. Clarke EB, Curtis JR, Luce JM, Levy M, Danis M, Nelson J, et al. Quality indicators for end-of-life care in the intensive care unit. Crit Care Med. 2003;31(9):2255–62.
19. Nelson JE, Mulkerin CM, Adams LL, Pronovost PJ. Improving comfort and communication in the ICU: a practical new tool for palliative care performance measurement and feedback. Qual Saf Health Care. 2006;15(4):264–71.
20. Penrod JD, Pronovost PJ, Livote EE, Puntillo KA, Walker AS, Wallenstein S, et al. Meeting standards of high-quality intensive care unit palliative care: clinical performance and predictors. Crit Care Med. 2012;40(4):1105–12.
21. Nelson JE, Bassett R, Boss RD, Brasel KJ, Campbell ML, Cortez TB, et al. Models for structuring a clinical initiative to enhance palliative care in the intensive care unit: a report from the IPAL-ICU Project (Improving Palliative Care in the ICU). Crit Care Med. 2010; 38(9):1765–72.
22. Goldsmith B, Dietrich J, Du Q, Morrison RS. Variability in access to hospital palliative care in the United States. J Palliat Med. 2008;11(8):1094–102.
23. Campbell ML, Guzman JA. A proactive approach to improve end-of-life care in a medical intensive care unit for patients with terminal dementia. Crit Care Med. 2004;32(9):1839–43.
24. Campbell ML, Guzman JA. Impact of a proactive approach to improve end-of-life care in a medical ICU. Chest. 2003;123(1):266–71.
25. Norton SA, Hogan LA, Holloway RG, Temkin-Greener H, Buckley MJ, Quill TE. Proactive palliative care in the medical intensive care unit: effects on length of stay for selected high-risk patients. Crit Care Med. 2007;35(6):1530–5.
26. Curtis JR, Cook DJ, Wall RJ, Angus DC, Bion J, Kacmarek R, et al. Intensive care unit quality improvement: a "how-to" guide for the interdisciplinary team. Crit Care Med. 2006;34(1):211–8.
27. Lilly CM, De Meo DL, Sonna LA, Haley KJ, Massaro AF, Wallace RF, et al. An intensive communication intervention for the critically ill. Am J Med. 2000;109(6):469–75.
28. Lilly CM, Sonna LA, Haley KJ, Massaro AF. Intensive communication: four-year follow-up from a clinical practice study. Crit Care Med. 2003;31(5 Suppl):S394–9.
29. Torke AM, Petronio S, Sachs GA, Helft PR, Purnell C. A conceptual model of the role of communication in surrogate decision making for hospitalized adults. Patient Educ Couns. 2012;87(1):54–61.

30. Baile WF, Buckman R, Lenzi R, Glober G, Beale EA, Kudelka AP. SPIKES-A six-step protocol for delivering bad news: application to the patient with cancer. Oncologist. 2000;5(4):302–11.
31. Lautrette A, Darmon M, Megarbane B, Joly LM, Chevret S, Adrie C, et al. A communication strategy and brochure for relatives of patients dying in the ICU. N Engl J Med. 2007;356(5):469–78.
32. Pollak KI, Arnold RM, Jeffreys AS, Alexander SC, Olsen MK, Abernethy AP, et al. Oncologist communication about emotion during visits with patients with advanced cancer. J Clin Oncol. 2007;25(36):5748–52.
33. Gay EB, Pronovost PJ, Bassett RD, Nelson JE. The intensive care unit family meeting: making it happen. J Crit Care. 2009;24(4):629.e1–12.
34. Davidson JE, Powers K, Hedayat KM, Tieszen M, Kon AA, Shepard E, et al. Clinical practice guidelines for support of the family in the patient-centered intensive care unit: American College of Critical Care Medicine Task Force 2004–2005. Crit Care Med. 2007;35(2):605–22.
35. Carlet J, Thijs LG, Antonelli M, Cassell J, Cox P, Hill N, et al. Challenges in end-of-life care in the ICU. Statement of the 5th International Consensus Conference in Critical Care: Brussels, Belgium, April 2003. Intensive Care Med. 2004;30(5):770–84.
36. Charles C, Gafni A, Whelan T. Decision-making in the physician-patient encounter: revisiting the shared treatment decision-making model. Soc Sci Med. 1999;49(5):651–61.
37. White DB, Braddock 3rd CH, Bereknyei S, Curtis JR. Toward shared decision making at the end of life in intensive care units: opportunities for improvement. Arch Intern Med. 2007;167(5):461–7.
38. Heyland DK, Cook DJ, Rocker GM, Dodek PM, Kutsogiannis DJ, Peters S, et al. Decision-making in the ICU: perspectives of the substitute decision-maker. Intensive Care Med. 2003;29(1):75–82.
39. Anderson WG, Arnold RM, Angus DC, Bryce CL. Passive decision-making preference is associated with anxiety and depression in relatives of patients in the intensive care unit. J Crit Care. 2009;24(2):249–54.
40. Johnson SK, Bautista CA, Hong SY, Weissfeld L, White DB. An empirical study of surrogates' preferred level of control over value-laden life support decisions in intensive care units. Am J Respir Crit Care Med. 2011;183(7):915–21.
41. Quill TE, Arnold RM, Platt F. "I wish things were different": expressing wishes in response to loss, futility, and unrealistic hopes. Ann Intern Med. 2001;135(7):551–5.
42. Stapleton RD, Engelberg RA, Wenrich MD, Goss CH, Curtis JR. Clinician statements and family satisfaction with family conferences in the intensive care unit. Crit Care Med. 2006;34(6):1679–85.
43. Scheunemann LP, Arnold RM, White DB. The facilitated values history: helping surrogates make authentic decisions for incapacitated patients with advanced illness. Am J Respir Crit Care Med. 2012;186(6):480–6.
44. Silveira MJ, Kim SY, Langa KM. Advance directives and outcomes of surrogate decision making before death. N Engl J Med. 2010;362(13):1211–8.
45. White DB, Arnold RM. The evolution of advance directives. JAMA. 2011;306(13):1485–6.
46. Tonelli MR. Pulling the plug on living wills. A critical analysis of advance directives. Chest. 1996;110(3):816–22.
47. Wendler D, Rid A. Systematic review: the effect on surrogates of making treatment decisions for others. Ann Intern Med. 2011;154(5):336–46.
48. Majesko A, Hong SY, Weissfeld L, White DB. Identifying family members who may struggle in the role of surrogate decision maker. Crit Care Med. 2012;40(8):2281–6.
49. Vig EK, Taylor JS, Starks H, Hopley EK, Fryer-Edwards K. Beyond substituted judgment: how surrogates navigate end-of-life decision-making. J Am Geriatr Soc. 2006;54(11):1688–93.
50. Curtis JR, Engelberg RA, Wenrich MD, Shannon SE, Treece PD, Rubenfeld GD. Missed opportunities during family conferences about end-of-life care in the intensive care unit. Am J Respir Crit Care Med. 2005;171(8):844–9.

51. McDonagh JR, Elliott TB, Engelberg RA, Treece PD, Shannon SE, Rubenfeld GD, et al. Family satisfaction with family conferences about end-of-life care in the intensive care unit: increased proportion of family speech is associated with increased satisfaction. Crit Care Med. 2004;32(7):1484–8.
52. Back AL, Arnold RM, Baile WF, Tulsky JA, Fryer-Edwards K. Approaching difficult communication tasks in oncology. CA Cancer J Clin. 2005;55(3):164–77.
53. White DB, Engelberg RA, Wenrich MD, Lo B, Curtis JR. Prognostication during physician-family discussions about limiting life support in intensive care units. Crit Care Med. 2007;35(2):442–8.
54. Evans LR, Boyd EA, Malvar G, Apatira L, Luce JM, Lo B, et al. Surrogate decision-makers' perspectives on discussing prognosis in the face of uncertainty. Am J Respir Crit Care Med. 2009;179(1):48–53.
55. Boyd EA, Lo B, Evans LR, Malvar G, Apatira L, Luce JM, et al. "It's not just what the doctor tells me:" factors that influence surrogate decision-makers' perceptions of prognosis. Crit Care Med. 2010;38(5):1270–5.
56. Apatira L, Boyd EA, Malvar G, Evans LR, Luce JM, Lo B, et al. Hope, truth, and preparing for death: perspectives of surrogate decision makers. Ann Intern Med. 2008;149(12): 861–8.
57. Schenker Y, White DB, Crowley-Matoka M, Dohan D, Tiver GA, Arnold RM. "It hurts to know... and it helps": exploring how surrogates in the ICU cope with prognostic information. J Palliat Med. 2013;16(3):243–9.
58. Schenker Y, Crowley-Matoka M, Dohan D, Tiver GA, Arnold RM, White DB. I don't want to be the one saying 'we should just let him die': intrapersonal tensions experienced by surrogate decision makers in the ICU. J Gen Intern Med. 2012;27(12):1657–65.
59. Zier LS, Sottile PD, Hong SY, Weissfield LA, White DB. Surrogate decision makers' interpretation of prognostic information: a mixed-methods study. Ann Intern Med. 2012;156(5):360–6.
60. Back A, Arnold R, Tulsky J. Mastering communication with seriously ill patients: balancing honesty with empathy and hope. Cambridge, UK/New York: Cambridge University Press; 2009.
61. Power TE, Swartzman LC, Robinson JW. Cognitive-emotional decision making (CEDM): a framework of patient medical decision making. Patient Educ Couns. 2011;83(2):163–9.
62. Loewenstein G. Hot-cold empathy gaps and medical decision making. Health Psychol. 2005;24(4 Suppl):S49–56.
63. Selph RB, Shiang J, Engelberg R, Curtis JR, White DB. Empathy and life support decisions in intensive care units. J Gen Intern Med. 2008;23(9):1311–7.
64. Puchalski C. Spirituality in health: the role of spirituality in critical care. Crit Care Clin. 2004;20(3):487–504, x.
65. Zier LS, Burack JH, Micco G, Chipman AK, Frank JA, Luce JM, et al. Doubt and belief in physicians' ability to prognosticate during critical illness: the perspective of surrogate decision makers. Crit Care Med. 2008;36(8):2341–7.
66. Zier LS, Burack JH, Micco G, Chipman AK, Frank JA, White DB. Surrogate decision makers' responses to physicians' predictions of medical futility. Chest. 2009;136(1):110–7.
67. Steinhauser KE, Voils CI, Clipp EC, Bosworth HB, Christakis NA, Tulsky JA. "Are you at peace?": one item to probe spiritual concerns at the end of life. Arch Intern Med. 2006;166(1):101–5.
68. Borneman T, Ferrell B, Puchalski CM. Evaluation of the FICA tool for spiritual assessment. J Pain Symptom Manage. 2010;40(2):163–73.
69. Anandarajah G, Hight E. Spirituality and medical practice: using the HOPE questions as a practical tool for spiritual assessment. Am Fam Physician. 2001;63(1):81–9.
70. Delisser HM. A practical approach to the family that expects a miracle. Chest. 2009; 135(6):1643–7.
71. Lo B, Kates LW, Ruston D, Arnold RM, Cohen CB, Puchalski CM, et al. Responding to requests regarding prayer and religious ceremonies by patients near the end of life and their families. J Palliat Med. 2003;6(3):409–15.

72. Azoulay E, Pochard F, Chevret S, Jourdain M, Bornstain C, Wernet A, et al. Impact of a family information leaflet on effectiveness of information provided to family members of intensive care unit patients: a multicenter, prospective, randomized, controlled trial. Am J Respir Crit Care Med. 2002;165(4):438–42.
73. Barry MJ. Health decision aids to facilitate shared decision making in office practice. Ann Intern Med. 2002;136(2):127–35.
74. Improving Palliative Care in the ICU. Available from: www.capc.org/ipal-icu/patient-family-resources. Accessed 14 Aug 2014.
75. Nelson JE, Cox CE, Hope AA, Carson SS. Chronic critical illness. Am J Respir Crit Care Med. 2010;182(4):446–54.
76. Carson SS, Vu M, Danis M, Camhi SL, Scheunemann LP, Cox CE, et al. Development and validation of a printed information brochure for families of chronically critically ill patients. Crit Care Med. 2012;40(1):73–8.
77. El-Jawahri A, Podgurski LM, Eichler AF, Plotkin SR, Temel JS, Mitchell SL, et al. Use of video to facilitate end-of-life discussions with patients with cancer: a randomized controlled trial. J Clin Oncol. 2010;28(2):305–10.
78. McCannon JB, O'Donnell WJ, Thompson BT, El-Jawahri A, Chang Y, Ananian L, et al. Augmenting communication and decision making in the intensive care unit with a cardiopulmonary resuscitation video decision support tool: a temporal intervention study. J Palliat Med. 2012;15(12):1382–7.
79. Beauchamp TL, Childress JF. Principles of biomedical ethics. 6th ed. New York: Oxford University Press; 1999.
80. Curtis JR, Vincent JL. Ethics and end-of-life care for adults in the intensive care unit. Lancet. 2010;376(9749):1347–53.
81. Asch DA, Faber-Langendoen K, Shea JA, Christakis NA. The sequence of withdrawing life-sustaining treatment from patients. Am J Med. 1999;107(2):153–6.
82. Gerstel E, Engelberg RA, Koepsell T, Curtis JR. Duration of withdrawal of life support in the intensive care unit and association with family satisfaction. Am J Respir Crit Care Med. 2008;178(8):798–804.
83. Seymour JE. Negotiating natural death in intensive care. Soc Sci Med. 2000;51(8):1241–52.
84. Christakis NA, Asch DA. Biases in how physicians choose to withdraw life support. Lancet. 1993;342(8872):642–6.
85. Cook DJ, Guyatt GH, Jaeschke R, Reeve J, Spanier A, King D, et al. Determinants in Canadian health care workers of the decision to withdraw life support from the critically ill. Canadian Critical Care Trials Group. JAMA. 1995;273(9):703–8.
86. Cook D, Rocker G, Marshall J, Sjokvist P, Dodek P, Griffith L, et al. Withdrawal of mechanical ventilation in anticipation of death in the intensive care unit. N Engl J Med. 2003;349(12):1123–32.
87. Vincent JL. Cultural differences in end-of-life care. Crit Care Med. 2001;29(2 Suppl):N52–5.
88. Mebane EW, Oman RF, Kroonen LT, Goldstein MK. The influence of physician race, age, and gender on physician attitudes toward advance care directives and preferences for end-of-life decision-making. J Am Geriatr Soc. 1999;47(5):579–91.
89. Fassier T, Azoulay E. Conflicts and communication gaps in the intensive care unit. Curr Opin Crit Care. 2010;16(6):654–65.
90. Schneiderman LJ. Defining medical futility and improving medical care. J Bioeth Inq. 2011;8(2):123–31.
91. Goold SD, Williams B, Arnold RM. Conflicts regarding decisions to limit treatment: a differential diagnosis. JAMA. 2000;283(7):909–14.
92. Emmanuell L, von Gunten C, Ferris F. Module 9: Medical Futility. The Education in Palliative and End-of-Life Care (EPEC) Curriculum; 1999, 2003.
93. Rubin SB. If we think it's futile, can't we just say no? HEC Forum. 2007;19(1):45–65.
94. O'Mahony S, McHugh M, Zallman L, Selwyn P. Ventilator withdrawal: procedures and outcomes. Report of a collaboration between a critical care division and a palliative care service. J Pain Symptom Manage. 2003;26(4):954–61.

95. Lewis JP, Ho KM, Webb SA. Outcome of patients who have therapy withheld or withdrawn in ICU. Anaesth Intensive Care. 2007;35(3):387–92.
96. Turner JS, Briggs SJ, Springhorn HE, Potgieter PD. Patients' recollection of intensive care unit experience. Crit Care Med. 1990;18(9):966–8.
97. Billings JA. Humane terminal extubation reconsidered: the role for preemptive analgesia and sedation. Crit Care Med. 2012;40(2):625–30.
98. Truog RD, Brock DW, White DB. Should patients receive general anesthesia prior to extubation at the end of life? Crit Care Med. 2012;40(2):631–3.
99. Faber-Langendoen K. The clinical management of dying patients receiving mechanical ventilation. A survey of physician practice. Chest. 1994;106(3):880–8.
100. Wilson WC, Smedira NG, Fink C, McDowell JA, Luce JM. Ordering and administration of sedatives and analgesics during the withholding and withdrawal of life support from critically ill patients. JAMA. 1992;267(7):949–53.
101. Campbell ML. Assessing respiratory distress when the patient cannot report dyspnea. Nurs Clin North Am. 2010;45(3):363–73.
102. Chanques G, Payen JF, Mercier G, de Lattre S, Viel E, Jung B, et al. Assessing pain in non-intubated critically ill patients unable to self report: an adaptation of the Behavioral Pain Scale. Intensive Care Med. 2009;35(12):2060–7.
103. Truog RD, Burns JP, Mitchell C, Johnson J, Robinson W. Pharmacologic paralysis and withdrawal of mechanical ventilation at the end of life. N Engl J Med. 2000;342(7):508–11.
104. Wee BL, Coleman PG, Hillier R, Holgate SH. The sound of death rattle II: how do relatives interpret the sound? Palliat Med. 2006;20(3):177–81.
105. Wee BL, Coleman PG, Hillier R, Holgate SH. The sound of death rattle I: are relatives distressed by hearing this sound? Palliat Med. 2006;20(3):171–5.
106. Erstad BL, Puntillo K, Gilbert HC, Grap MJ, Li D, Medina J, et al. Pain management principles in the critically ill. Chest. 2009;135(4):1075–86.
107. Campbell ML. Terminal dyspnea and respiratory distress. Crit Care Clin. 2004;20(3):403–17, viii–ix.

Chapter 6
Cardiac Patients at End of Life in the Emergency Department

Derrick Lowery and Christopher R. Carpenter

Abstract Patients with symptoms related to end-stage cardiac disease, including coronary artery disease and heart failure, commonly present to the emergency department. Emergency healthcare providers' understanding of disease trajectory aligned with patient and family goals of care facilitates an optimal approach to individualized management. This review highlights existing cardiac-specific prognostic instruments within the context of contemporary emergency and palliative care models, emphasizing the benefits of efficient and pre-organized communication strategies.

Keywords Heart failure • Emergency medicine • End-of-life • Palliative care • Prognosis • Medical device

Key Points

- Due to the aging of the population and the high prevalence of cardiovascular disease, ED providers are increasingly confronted with patients who have acute or chronic life-threatening or pre-terminal cardiac conditions.
- Introducing palliative care concepts early while continuing life-prolonging active treatment improves patient and family satisfaction, decreases costs, reduces time in intensive care units, and ensures that the care trajectory initiated in the ED aligns with patient wishes.
- Prognostication for heart failure patients is difficult because illness trajectory is highly variable. Therefore, emphasis on function as it relates to the patient's life goals may be preferred when discussing care plan options.

D. Lowery, MD (✉)
Hospice and Palliative Medicine Fellow, Department of Internal Medicine,
Division of Geriatric Medicine and Gerontology, Emory University, Atlanta, GA, USA
e-mail: derrick.lowery@emory.edu

C.R. Carpenter, MD, MSc
Department of Emergency Medicine, Washington University in St. Louis School of Medicine and Barnes-Jewish Hospital, St. Louis, MO, USA

S.J. Goodlin, M.W. Rich (eds.), *End-of-Life Care in Cardiovascular Disease*,
DOI 10.1007/978-1-4471-6521-7_6

- Palliative emergency medicine experts advocate systematic palliative assessments in conjunction with a formal organized approach to communication with patients and families.
- Emergency medicine physicians must be familiar with common cardiac devices, including ICDs and LVADs, and be prepared to discuss deactivation of these devices with patients and families in the event of immediately life-threatening or terminal illness.

Emergency Department (ED) and Emergency Medical Services providers often encounter patients with cardiac pathology who are facing life-changing, life-threatening, or life-ending illnesses. These situations include acute coronary syndrome, cardiac arrest, heart failure (HF), arrhythmias, and complications of medical devices and therapies used to treat cardiac diseases. With advances in therapy, patients with heart disease, including coronary heart disease and myocardial infarction (MI), are surviving longer and subsequently developing HF and arrhythmias. The prevalence of HF in the United States (U.S.) is over five million, and there are an estimated 650,000 new cases per year [1]. While overall mortality for these patients is improving, mortality remains high for patients admitted to the hospital for HF with 42.3 % dying within 5 years of hospitalization. Further, HF patients have a substantial rate of readmission, with 24.5 % of patients being rehospitalized within 30 days of discharge [2].

Historically, ED management prioritized life-prolonging care for patients with serious and potentially life-threatening illnesses. However, it is now necessary for ED clinicians to expand beyond simply providing life-prolonging care to become proficient at providing palliative care when appropriate. Palliative care includes symptom management of the dying patient or the patient facing a decline of function. Palliative care also emphasizes clarification of patients' goals of care, psychological support, spiritual support, social support, and assistance with resource identification. The benefits to integrating palliative care within the ED include improved patient satisfaction, cost savings, reduced length of stay, and less use of intensive care [3].

Assessment of Prognosis and Identification of Cardiac Patients for Whom Palliative Intervention Is Appropriate

Care of the patient approaching the end of life can be a difficult task in the ED setting where rapid decision-making is the norm [1]. Palliative intervention for cardiac patients with severe disease, co-morbid terminal illness, or acute functional decline is often delayed by ED physicians because they believe that these interventions are beyond their scope of practice or that appropriate palliative interventions will be introduced after the patient is admitted. However, early recognition of the need for palliative interventions ensures that the care trajectory and course of hospitalization aligns with the patient's wishes. In addition, by simultaneously combining the palliative model of care with life-prolonging therapy early in the disease course, a more natural transition to exclusive palliative care can occur later when life-prolonging therapies are no longer an option [4].

Barriers to the implementation of palliative care in the ED include: a perception that palliative discussions are excessively time consuming, lack of coordination between primary care doctors and ED doctors, failure to recognize terminal status, lack of knowledge of palliative care options, lack of resources, medicolegal concerns, lack of access to complete medical records, lack of access to palliative care teams, and an ED culture which favors reflexive use of extensive technology and intensive therapies [5, 6].

Prior to implementing palliative interventions, the ED provider must first identify and recognize that a patient is on a dying trajectory. Seven categories of death or near death scenarios in ED patients provide a framework for aiding physicians in guiding the patient and family to select a care plan in line with realistic goals [7]. These categories are:

1. Dead on arrival
2. Prehospital resuscitation with subsequent death in the ED
3. Prehospital resuscitation with survival until admission
4. Terminally ill and arrives at the ED. These patients, their families, and their primary care doctor have all formally recognized and accepted the terminal nature of the underlying disease.
5. Frail and hovering near death. These patients may be similar to the terminally ill patients identified above, but are distinctly different in that the patient or family have not formally recognized, acknowledged, and/or accepted the terminal nature of the patient's condition.
6. Alive and interactive on arrival but has a cardiac or pulmonary arrest in the ED.
7. Potentially preventable death by omission or commission; i.e., the patient's death results from actions or inactions taken by the care team in the hospital.

Understanding the functional trajectory of patients with life-limiting illnesses (Fig. 6.1) also empowers ED physicians to guide patients and families towards individualized, goal-directed courses of action. Lunney et al. have proposed four functional trajectories [8]:

1. **Sudden death** – a patient without any known history of underlying serious disease who is fully functional and has an unexpected catastrophic event that leads to death (acute MI, aortic dissection, arrhythmia, pulmonary edema, pericardial tamponade).
2. **Disseminated cancer** – a patient with cancer who has maintained adequate function until the final 2–4 months of life when tumor burden causes a rapid functional decline.
3. **Organ failure trajectory** – a patient with an underlying disease process such as chronic obstructive pulmonary disease (COPD), chronic HF, or cirrhosis that is

characterized by intermittent exacerbations associated with acute decreases in function that improve or resolve after an intervention but with a progressive decline in function and a downward-sloping sinusoidal course.

4. **Frailty** – a patient with a slowly progressing fatal disease such as dementia in which function declines gradually over a long period of time (usually years).

In most cases, it is difficult for ED physicians to provide a specific prognosis to patients and their families. This challenge may lead to avoidance of the topic. However, attempts at prognostication for patients at the end of life may allow patients and their families the opportunity to set realistic expectations. Unfortunately, patients with HF often have unpredictable trajectories that do not fit within the above models. Therefore, ED physicians may appropriately avoid specific time frames and instead focus on anticipated declines in function and how this may impact the patient's goals [9].

Despite the variation in death trajectory for HF patients, tools designed to aid in prognostication have been developed. The New York Heart Association (NYHA) functional classification categorizes patients according to symptom severity and functional limitations.

New York Heart Association Functional Classification

- Class 1: No limitation of physical activity. Ordinary physical activity does not cause undue fatigue or dyspnea.
- Class 2: Slight limitation of physical activity. Comfortable at rest, but ordinary physical activity results in fatigue or dyspnea.
- Class 3: Marked limitation of physical activity. Comfortable at rest, but less than ordinary physical activity, such as walking short distances, causes fatigue or dyspnea.
- Class 4: Severe limitation in activities. Unable to carry on any physical activity without symptoms. Symptoms may be present at rest.

Although the NYHA classification does not specifically address predicted time until death, patients with class 4 symptoms despite optimal medical therapy have a median survival of less than 6 months. Thus, NYHA class can be used in conjunction with the Fig. 6.1 death trajectories for framing a discussion regarding care [10]. However, many HF patients do not progress sequentially through the NYHA stages prior to death, and many class 1–3 patients die suddenly due to an arrhythmia or other acute deterioration in their condition [11].

Another prognostication instrument is the Seattle Heart Failure Model (available at www.SeattleHeartFailureModel.org), which provides estimates of mean life expectancy and survival rates at 1, 2, and 5 years, as well as information on how various medical therapies or devices alter life expectancy [12].

These tools aid physicians in leading a discussion with patients and their families. The presence of the patient in the ED indicates that the patient or fam-

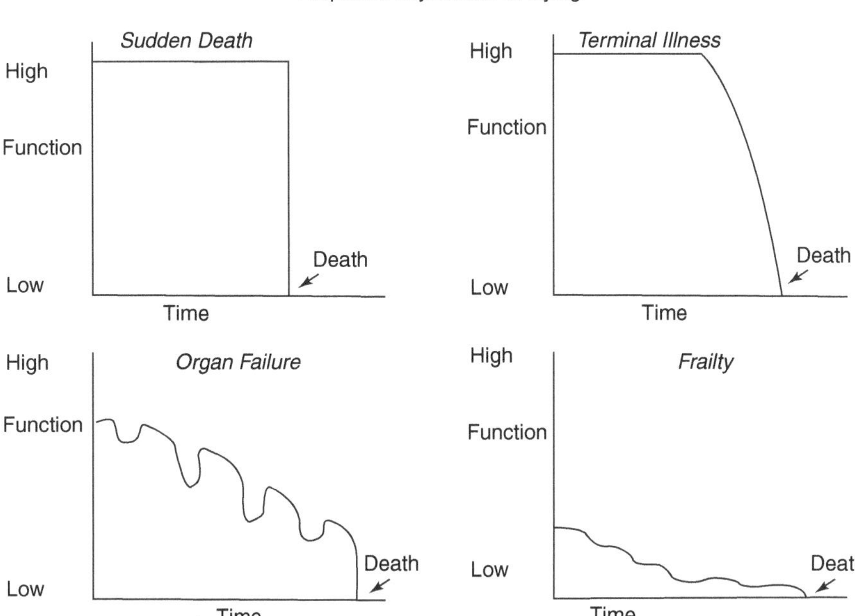

Fig. 6.1 Trajectories of dying (From Lunney et al. [8]. Reprinted with permission from John Wiley and Sons)

ily is in distress from physical symptoms, emotional duress, or unmet social/spiritual needs. A discussion of prognosis does not reduce or eliminate distress, but can be used to align patient and family expectations with realistically achievable objectives followed by goal-directed implementation of a management plan [13]. Regardless of prognosis, the emphasis of care focuses on alleviating whatever factors contributed to the patient presenting to the ED. Listening carefully to the patient and family may reveal that the prognosis is already understood, but that there are acute stressors which the astute ED physician can effectively alleviate.

Palliative Assessment in the ED

In patients with significant cardiac disease that is life changing or potentially life ending, the ED provider conducts a structured palliative needs assessment to identify the totality of palliative care needs. Two proposed models for accomplishing this task are the Focused Assessment (ABCD) [13], which is appropriate for unstable or critically ill patients, and the expanded Needs at the End of life Screening Tool (NEST), which can be performed by an interdisciplinary team [14].

The ABCD instrument allows for a rapid palliative assessment and can be used in situations where time is limited (i.e. cardiac arrest, septic shock) and may be performed concurrently with other interventions by a non-physician member of the care team. Although the mnemonic fits neatly into alphabetical order, the questions may be better addressed in reverse order.

A. Advance Directive or advance care plan – does one exist? Is it available for review?
B. Better symptom control – how can the care team address distressing symptoms?
C. Caregivers – who are the key caregivers, and are they present or available?
D. Decisional capacity – does the patient have the capacity to make healthcare decisions? Even in the setting of profound pain or dyspnea, patients deserve the opportunity to express basic wishes concerning their healthcare.

The NEST screening tool (Table 6.1) addresses four domains of care for a patient with possible palliative needs. The tool has been studied in the ED and identifies a significant number of such needs that are often unrecognized without a structured approach. However, there has not been a direct comparison of the NEST with a non-formalized physician 'gestalt' assessment [15] (Table 6.1).

Table 6.1 The NEST assessment

Needs	Are there social needs (including financial and caregiver issues) that should guide patient disposition?	1. How much of a financial hardship is your illness for your family?
		2. How much trouble do you have accessing the medical care you need?
		3. How often is there someone to confide in?
		4. How much help do you need with things like getting meals or getting to the doctor?
Existential	Does the patient have existential needs (i.e., concerns about meaning/ spiritual/faith/forgiveness) that contribute to distress?	1. How much does this illness seem senseless and meaningless?
		2. How much does religious belief or your spiritual life contribute to your sense of purpose?
		3. How much have you settled your relationships with the people close to you?
		4. Since your illness, how much do you live your life with a special sense of purpose?
Symptom matters	What physical or mental symptoms led to this Emergency Department visit?	1. How much do you suffer from physical symptoms such as pain, shortness of breath, fatigue, bowel or urinary problems?
		2. How often do you feel confused, anxious, or depressed?
Therapeutic	What should the therapeutic goals for this Emergency Department visit be?	1. How much do you feel your doctors and nurses respect you as an individual?
		2. How clear is the information from the medical team about what to expect regarding your illness?
		3. How much do you feel that the medical care you are getting fits with your goals?

Communication

Effective communication is fundamental to the practice of emergency medicine, and it is particularly critical when providing care to patients with grim prognoses or when conveying life-changing news to families. Efficient and effective communication is challenging because of the busy nature of the ED, the lack of an established relationship with patients and their families, insufficient information regarding previous discussions between family and primary care physicians, discontinuous or inconsistent access to relevant clinical data, and frequent interruptions common to the ED work environment [11].

Despite these challenges, practitioners who are dual trained in palliative medicine and emergency medicine have promoted employing a standardized approach to communication. These techniques assist ED providers in caring for the cardiac patient whose condition deteriorates or who suffers a cardiac arrest or a life-changing co-morbid event. These situations provide an opportunity for the physician to work with the patient and family to consider the goals of care and how these goals change as the patient's condition evolves.

The six step SPIKES protocol has been studied as a means for optimizing the success of delivering bad news and has been advocated as an ED-appropriate tool [16].

- **S** – Setting up the interview: arrange for privacy, involve appropriate family members, sit down, manage time constraints and avoid interruptions by notifying other ED staff that you will be having this discussion and request that they minimize interruptions. Consider mentally rehearsing your message, and prepare possible answers to likely questions. Make a connection with the patient by establishing rapport and being mindful of body language. Sit down if possible, make eye contact, maintain an open and forward body position, and consider appropriate physical touch.
- **P** – Perception: Assess the patient's perception of the current clinical situation. Use open-ended questions, correct any misinformation, listen for evidence of denial or unrealistic expectations. Starting the conversation with a phrase such as "Tell me what you understand about your condition" allows the conversation to proceed in a patient-centered manner.
- **I** – Invitation: Obtain the patient's invitation / permission to share information. Determine who should receive information and how much information should be shared. This step is important particularly in cultures where patients wish to avoid being informed of any bad news and prefer to have that information shared with a surrogate. It is reasonable for a patient to request NOT to be informed of bad news when that patient appoints a surrogate family member or care provider to discuss the clinical plan.
- **K** – Knowledge: Provide information to the patient. Warn listeners prior to delivering any bad news. Use simple, non-jargon filled language. Avoid excessive bluntness. Give information in small chunks. Use of a "warning shot" such as "I'm afraid I have bad news" prepares the listener for the delivery of unpleasant news. Do not minimize the seriousness of the news in an attempt to provide comfort as this may elicit confusion. If the prognosis is known with a reasonable

degree of certainty, it may be appropriate to discuss it if the conversion will assist in planning further care.

- **E**- Emotions: Address the patient's emotions with empathetic responses. Observe the patient's emotional responses. Identify the emotions; if needed clarify with open-ended questions. Identify the reason for the emotion. Allow time for the patient/family to express emotions and respond with empathetic statements. Be prepared for a wide range of emotional responses including denial, anger, sadness, guilt or even relief, and validate that the emotional response is not inappropriate. Phrases such as "I see that this is painful for you" and "I wish this was not happening to your loved one" demonstrate compassion. Offering support such as calling a chaplain or other family member may be helpful. Having tissue in the room is helpful, but directly offering tissue to a person who is crying may be perceived as a request that the patient stop crying and "clean themselves up". It may be preferable to wait for the patient to take the tissue when ready to use it.
- **S** – Strategy / Summary: Discuss next steps and verify understanding. It is important to invite the patient or family member to summarize the information that you have delivered to ensure that there are no misunderstandings. A proposed phraseology to accomplish this is "I know this has been overwhelming but I want to make sure that everything is understood before we move on. Please tell me what you understand about all of this?"

ED staff are often tasked with notifying a family that their loved one has died. The Education in Palliative and End of Life Care Emergency Medicine Curriculum advocates use of the GRIEV_ING mnemonic for death disclosures [17]. This tool encompasses components similar to the SPIKES protocol but has been specifically designed for the ED setting. Clinicians trained in its use have been shown to have improved notification skills. Familiarity with this standardized protocol for death disclosure provides the busy ED physician with a tool to aid in performing this task after a cardiac arrest or other fatal event when emotions may run high [18].

- **G** – Gather: gather the family; ensure that all members are present.
- **R** – Resources: call for support resources to assist the family with their grief (i.e. chaplain services, family, friends).
- **I** – Identify: identify yourself, identify the deceased or injured patient by name, and identify the state of knowledge of the family relative to the events of the day.
- **E** – Educate: briefly educate the family as to the events that have occurred in the ED; educate them about the current state of their loved one.
- **V** – Verify: verify that their family member has died. Be Clear! Use the words "dead" or "died."
- _ – Space: Give the family personal space and time for an emotional moment; allow the family time to absorb the information.
- **I** – Inquire: ask if there are any questions, and answer them all.
- **N** – Nuts and Bolts: inquire about organ donation, funeral services, and personal belongings. Offer the family the opportunity to view the body.
- **G** – Give: give them your card and contact information. Offer to answer any questions that may arise later. Always return their call.

In the last decade, there has been a shift in views about whether to allow family to be present during active resuscitation attempts in the setting of cardiac arrest [19]. Witnessing the resuscitation may be helpful in the grieving process for the patient's loved ones [20]. Conversely, some providers express concerns that allowing families to witness an attempt at resuscitation may interfere with clinical care, expose physicians to additional legal burdens, cause undue emotional distress to families, or add to the emotional tensions experienced by staff. However, none of these concerns have been validated in studies of EDs that have implemented family presence policies during resuscitation [19].

Ideally, a protocol should be in place prior to offering family members the option of being present during resuscitative efforts. The protocol should include assignment of a staff member responsible for providing emotional support to the family. This individual should assist the family in finding a physical place to be during the resuscitation, be available to answer questions about medical interventions, and serve as a liaison to the medical team. This staff member should be free from any other clinical duties during the resuscitation. The protocol may empower the physician-in-charge to determine whether family presence would be appropriate in individual circumstances. The physician should have the opportunity to make this decision outside of family presence and before they are invited to witness the resuscitation. Prior to inviting family into the room, inform staff that family will be present to ensure appropriate attention to patient modesty and dignity.

When the family is present during a resuscitation attempt, the decision to stop cardiopulmonary resuscitation and pronounce the patient dead may feel uncomfortable for some providers. The EPEC-EM curriculum advocates a six-step approach to stopping CPR, including sample language as follows:

1. Deliver a warning: "We have a seventy-year-old female who suffered a cardiac arrest secondary to drowning."
2. Recap events: "The resuscitation has been in progress for forty minutes from pre-hospital to now with successful airway control, effective chest compressions, fluid resuscitation with two boluses of warmed saline, and correction of mild hypothermia from a core temperature of 34.6 to 37 degrees Celsius. Patient has been asystolic since arrival to the ED. Ultrasound shows no cardiac activity." This recap is primarily directed towards soliciting input from other medical team members. It is important that the medical jargon used during this recap be later explained to the family members present in terms that they can understand.
3. Allow the team to give suggestions: "Does anybody have any other suggestions for interventions that might help this patient?"
4. Explicit statement of team comfort about cessation: "Is everyone comfortable with stopping resuscitation?"
5. Pronounce death: "With team consensus, death is pronounced at 14:55."
6. Thank the team, acknowledge difficulty, and encourage processing: "Thank you, everyone. This was a very difficult case, and I appreciate everyone's efforts. Please take a moment to reflect in whatever way feels appropriate."

Cardiac Devices at End of Life

The presence of intracardiac devices such as pacemakers, implantable cardioverter-defibrillators (ICDs), and left ventricular assist devices (LVADs) adds complexity for the physician caring for a patient facing the end of life [21]. Burdens associated with these devices at the end of life include the delivery of painful shocks to a patient during the dying process and ethical concerns related to turning off such a device. Further, some devices may make it challenging to determine when a patient has actually died. For example, an LVAD may produce mechanical forward cardiac flow even after the patient has died.

Ideally, the conversation about when to deactivate or remove an artificial cardiac device should take place prior to insertion [22]. At the time of implantation, when the patient has a satisfactory quality of life, avoidance of sudden cardiac death is a reasonable and appropriate goal. However, as the patient's health deteriorates and quality of life declines, a painless cardiac death may be acceptable and in line with the patient's wishes. Further, the principle of patient autonomy holds that patients have the right to refuse or discontinue any therapies, including devices that are no longer consistent with their goals of care. An early frank discussion of these issues with the patient is essential, as studies demonstrate that patients are more comfortable engaging in these conversations prior to facing the end of life [23].

Physicians are often reluctant to engage in discussions regarding device deactivation [24, 25]. In one study, 278 patients with defibrillators were surveyed, and although half had advance directives, only 1 % made mention of plans for their device at the end of life [26]. This low rate of mentioning deactivation has been confirmed in other studies [27]. Other research has demonstrated that most patients are comfortable discussing deactivation. In one study, 71 % of patients with ICDs indicated that they would choose deactivation in at least one of several hypothetical scenarios [28]. These findings imply a failure of physicians to lead the conversation, clarify patient goals, and document patient preferences.

A simple approach for ED physicians assessing patients with ICDs is to ask whether they have documented the circumstances, if any, in which they would want their ICD turned off. Encouraging healthy patients with ICDs to have this discussion with their cardiologist, primary care doctor, and family and to document their wishes may prevent subsequent uncertainty. The ED physician can also advise the patient that "If at some future date, you decide you no longer want the device, it can easily be disabled" [21]. This phrase serves to empower patients by ensuring awareness of the deactivation option, while providing an opportunity to contemplate their wishes.

In the event that a patient with an ICD is expected to die in the ED setting after the patient's goals have been clarified, it is important for the ED physician to be familiar with magnet-deactivation of the device. Placing a magnet over the device should prevent it from firing. Importantly, this does not disable the device's pacemaker functions. Instead, it causes the device to revert to asynchronous pacing at a preset rate. Failure of the magnet to have the desired effect may be caused by poor positioning or patient body habitus, in which case the magnet and/or patient should be repositioned.

ED physicians should specifically address deactivation when referring patients with implanted devices to a hospice setting. Many hospices do not have formal policies for dealing with implanted devices, and failure to address deactivation has led to multiple cases of shocks being delivered at the end of life. In one study, 8 % of patients with ICDs received shocks in the final minutes of life [28]. These shocks are unnecessarily painful to the dying patient, provoke anxiety, and may (though not always) be contrary to the expressed wishes of the patient entering into hospice care, which focuses primarily on patient comfort. Further, repetitive shocks may cause emotional trauma to family members and hospice staff who witness these events and are unable to terminate them [29].

Some physicians are opposed to deactivation of an ICD because they believe that it equates to physician assisted suicide [25]. The Heart Rhythm Society issued a consensus document that dispels this concern and states that "carrying out a request to withdraw life-sustaining treatment is neither physician assisted suicide (PAS) nor euthanasia… the right to refuse or request the withdrawal of a treatment is a personal right of the patient" [22]. A subtle yet key distinction that may assist a physician troubled by the ethics of such a decision is that PAS or euthanasia involves active intervention undertaken by a physician that brings about the death of a patient, whereas deactivation of a device is the discontinuation or removal of a medical intervention in order to respect the autonomy of a patient. If a physician continues to object to deactivation on moral, religious, or ethical grounds, it is recommended that he or she collaborate with or transfer care to a physician who is willing to act in accordance with the patient's wishes. It is important that the patient's autonomy is protected by ensuring that the patient is not experiencing undue pressure, does not have underlying depression of sufficient severity to impede rational decision making, is not actively suicidal, and understands the ramifications of device deactivation. A palliative care provider or psychiatrist may be needed to assist in clarifying the patient's goals of care and to ensure preservation of patient autonomy.

LVADs are another device that emergency medicine (EM) physicians must be prepared to deal with when caring for patients in the ED. Historically, LVADs were primarily used in hospitalized patients as a "bridge-to-transplant". However, advances in technology coupled with the fact that the number of patients with advanced HF has increased while the number of available donor hearts has remained stable has led to a shift in the role of the LVAD. Currently, many patients with LVADs have the devices implanted without any further interventions planned, a strategy referred to as a "destination therapy" (LVAD-DT). In 2012, 42 % of patients received LVADs as a bridge-to-transplant, 6 % were implanted for bridge-to-transplant candidacy, and 52 % were intended as destination therapy [30]. Increasingly, EM providers will be seeing patients near death with these devices. In the LVAD-DT patient, eventual death without transplant is the expected outcome. Indeed, "death is a guaranteed outcome; it is a natural part of life and clinicians who view death as a failure of medical therapies and technology will have difficulty engaging in discussions around future plans in general and device deactivation in particular" [31]. Ideally, the primary care team proactively prepares the LVAD patient and their family for the possibility of death with the device in place. However, in many situations the ED physician may become responsible for end-of-life discussions and decisions [32]. Assistance with these discussions should be aided by a member of the LVAD team whenever possible; however, many circumstances will require the ED physician to be equipped to answer patient and family questions and lead the discussion without the benefit of this resource.

Discontinuation of the LVAD creates ethical concerns for some physicians. Turning off an LVAD will result in death within 20 min if the patient is completely dependent on the device as are many patients. In contrast, although turning off a preventative device such as an ICD may contribute to the patient's future death, it is unlikely to result in a rapid demise. While this reality may cause unease among physicians, it is important to respect the right of the patient to accept or refuse medical treatment. Specifically, once a patient's goals of care are clarified, if these goals are being hindered by an intervention, including an LVAD, it is the duty of the physician to withhold or withdraw the intervention (see Fig. 6.2). Of note, ethical principles and legal precedent have established that in the end of life scenario there is no distinction between withholding and withdrawing an intervention [22, 33]. A further ethical consideration is the moral obligation that the physician has to both current and future patients. In other words, if a physician fails to respect an individual patient's wishes to withdraw LVAD support, it may contribute to a climate wherein future patients considering destination LVADs may fear losing the right to self-determination should their clinical condition change and they desire a more natural death. As a result, patients may choose to decline an LVAD as DT despite its potential for increasing longevity and improving quality of life [31, 34–36].

Conclusions

ED healthcare providers increasingly encounter patients with cardiac conditions that are potentially life-limiting or life-ending. ED physicians, nurses, and other staff directly involved in patient care should be prepared to address the palliative needs of cardiac patients, including clarification of goals of care and basic management of cardiac devices. Providers should use an organized approach to identify palliative needs in order to ensure that the care plan is in alignment with patient and family preferences and goals.

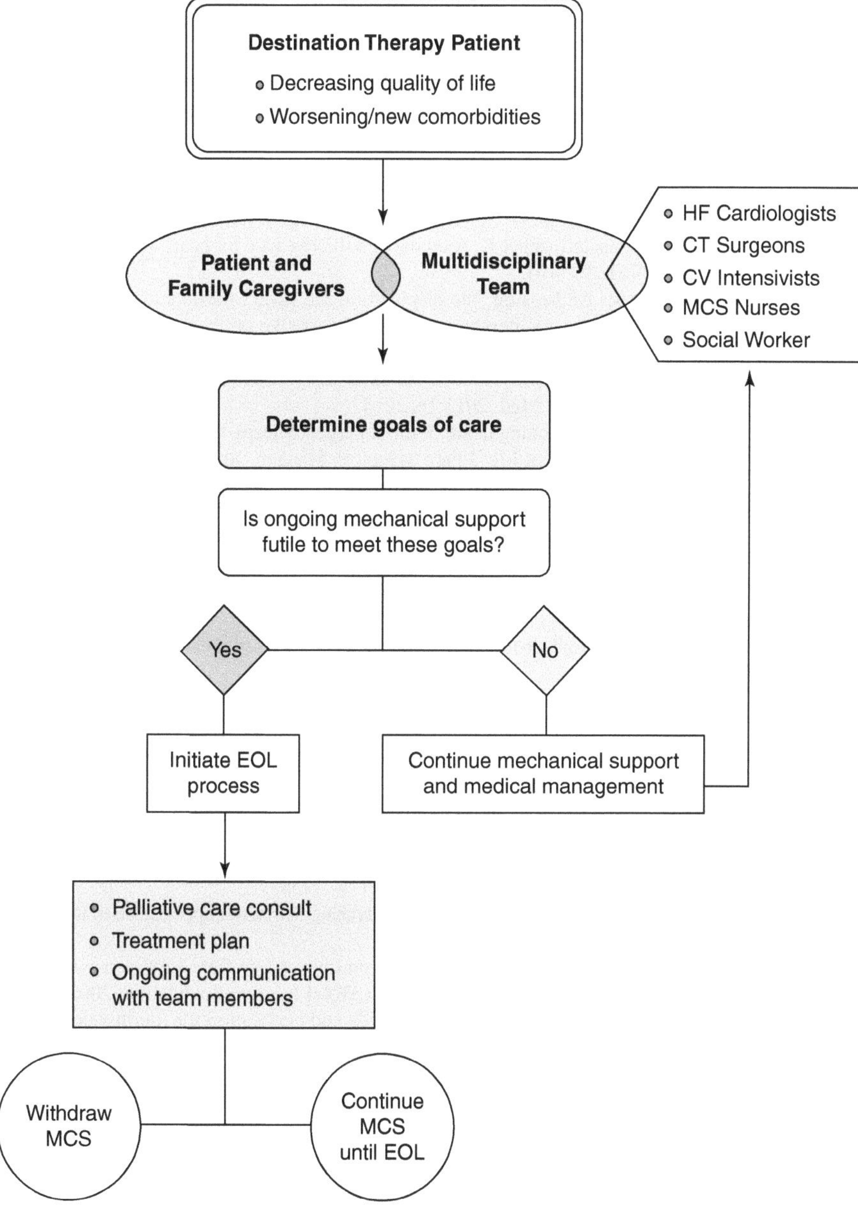

Fig. 6.2 Flowchart depicting proposed algorithm for approaching patients with mechanical circulatory support at end-of-life using a multi-disciplinary team. *HF* heart failure, *CT* cardiothoracic, *CV* cardiovascular, *MCS* mechanical circulatory support, *EOL* end-of-life (From Brush et al. [32])

References

1. Silvers S, et al. Critical issues in the evaluation and management of adult patients presenting to the ED with acute heart failure syndromes. Ann Emerg Med. 2007;49(5):627–36.
2. Go AS, et al. Heart disease and stroke statistics – 2013 update: a report from the American Heart Association. Circulation. 2013;127:e6–245.
3. Rosenberg M, Lamba S, Misra S. Palliative medicine and geriatric emergency care: challenges, opportunities, and basic principles. Clin Geriatr Med. 2013;29(1):1–29.
4. Hupcey JE, Penrod J, Fenstermacher K. A model of palliative care for heart failure. Am J Hosp Palliat Care. 2009;26(5):399–404.
5. Momen NC, Barclay SI. Addressing 'the elephant on the table': barriers to end of life care conversations in heart failure – a literature review and narrative synthesis. Curr Opin Support Palliat Care. 2011;5(4):312–6.
6. Lamba S, et al. Palliative care provision in the emergency department: barriers reported by emergency physicians. J Palliat Med. 2013;16(2):143–7.
7. Chan G. Trajectories of approaching death in the Emergency Department: clinician narratives of patient transitions to the end of life. J Pain Symptom Manage. 2011;42(6):864–81.
8. Lunney J, Lynn J, Hogan C. Profiles of Older Medicare decedents. J Am Geriatr Soc. 2002;50:1108–12.
9. Gott M, et al. Dying trajectories in heart failure. Palliat Med. 2007;21(2):95–9.
10. The Criteria Committee of the New York Heart Association. Nomenclature and criteria for diagnosis of diseases of the heart and great vessels. 9th ed. Boston: Little, Brown & Co; 1994. p. 253–6.
11. Friesinger GCC, Butler J. End-of-life care for elderly patients with heart failure. Clin Geriatr Med. 2000;16(3):663–75.
12. Nakayama M, Osaki S, Shimokawa H. Validation of mortality risk stratification models for cardiovascular disease. Am J Cardiol. 2011;108(3):391–6.
13. DeSandre PL, Quest T, editors. Palliative aspects of emergency care. New York: Oxford University Press; 2013.
14. Emanuel L, Alpert H, Emanuel E. Concise screening questions for clinical assessment of terminal care: the needs near the end-of-life care screening tool. J Palliat Care. 2001;4(4):465–74.
15. Grudzen CR, et al. Palliative care needs of seriously ill, older adults presenting to the emergency department. Acad Emerg Med. 2010;17(11):1253–7.
16. Baile W, et al. SPIKES – a six step protocol for delivering bad news: application to the patient with cancer. Oncologist. 2000;5(4):302–11.
17. Emmanuel L, editor. The Education for Physicians on End-of-Life Care-Emergency Medicine (EPEC-EM) curriculum. The EPEC Project: Robert Wood Johnson Foundation; 2008.
18. Hobgood C, et al. Griev_ing: death notification skills and application for fourth year medical students. Teach Learn Med. 2009;21(3):207–19.
19. Critchell CD, Marik PE. Should family members be present during cardiopulmonary resuscitation? A review of the literature. Am J Hosp Palliat Care. 2007;24(4):311–7.
20. Jabre P, Belpomme V, Azoulay E, et al. Family presence during cardiopulmonary resuscitation. N Engl J Med. 2013;368:1008–18.
21. Morgenweck CJ. Ethical considerations for discontinuing pacemakers and automatic implantable cardiac defibrillators at the end-of-life. Curr Opin Anaesthesiol. 2013;26(2):171–5.
22. Lampert R, et al. HRS Expert Consensus Statement on the Management of Cardiovascular Implantable Electronic Devices (CIEDs) in patients nearing end of life or requesting withdrawal of therapy. Heart Rhythm. 2010;7(7):1008–26.
23. Strachan PH, et al. Patients' perspectives on end-of-life issues and implantable cardioverter defibrillators. J Palliat Care. 2011;27(1):6–11.
24. Barclay S, et al. End-of-life care conversations with heart failure patients: a systematic literature review and narrative synthesis. Br J Gen Pract. 2011;61(582):e49–62.

25. Russo JE. Original research: deactivation of ICDs at the end of life: a systematic review of clinical practices and provider and patient attitudes. Am J Nurs. 2011;111(10):26–35.
26. Kirkpatrick JN, Kim AY. Ethical issues in heart failure: overview of an emerging need. Perspect Biol Med. 2006;49(1):1–9.
27. Kramer DB, et al. Ethical and legal views of physicians regarding deactivation of cardiac implantable electrical devices: a quantitative assessment. Heart Rhythm. 2010;7(11):1537–42.
28. Dodson J, et al. Preferences for ICD deactivation in the setting of advanced illness. American Geriatrics Society annual meeting, slides. 2012. http://www.americangeriatrics.org/files/documents/annual_meeting/2012/handouts/thursday/R0730-5308_John_A._Dodson.pdf.
29. Fromme E, et al. Adverse experiences with implantable defibrillators in Oregon hospices. Am J Hosp Palliat Care. 2011;28:304–9.
30. Kirklin JK, et al. The Fourth INTERMACS Annual Report: 4,000 implants and counting. J Heart Lung Transplant. 2012;31(2):117–26.
31. Matlock DD, Stevenson LW. Life-saving devices reach the end of life with heart failure. Prog Cardiovasc Dis. 2012;55(3):274–81.
32. Brush S, et al. End-of-life decision making and implementation in recipients of a destination left ventricular assist device. J Heart Lung Transplant. 2010;29(12):1337–41.
33. Rizzieri AG, et al. Ethical challenges with the left ventricular assist device as a destination therapy. Philos Ethics Humanit Med. 2008;3:20.
34. Pellegrino ED. Decisions to withdraw life-sustaining treatment: a moral algorithm. JAMA. 2000;283(8):1065–7.
35. Powell TP, Oz MC. Discontinuing the LVAD: ethical considerations. Ann Thorac Surg. 1997;63(5):1223–4.
36. Dudzinski DM. Ethics guidelines for destination therapy. Ann Thorac Surg. 2006;81(4):1185–8.

Chapter 7
Management of Hospitalized Patients with Unexpected Cardiopulmonary Arrest

Michael G. Dickinson, Christopher M. Meeusen, and Daniel L. Maison

Abstract The optimal management of an in-hospital cardiac arrest goes beyond the measures employed during the resuscitation. Clinicians should possess skills to deliver bad news and to guide decision making discussions. Scores are available and should be utilized to aid in prognostication to better guide these discussions. Prediction of neurologic outcome is challenging in the era of therapeutic hypothermia but the wealth of literature available provides a guide to understanding which patients may survive and which patients are unlikely to do so. Finally, using risk scores to guide code status discussions and systems-based patient safety measures can reduce the number of in-hospital cardiac arrests.

Keywords Cardiac arrest • Hypothermia • Neurologic outcome • Prognostication • Resuscitation

Key Points

- Hospitalized patients who suffer an unexpected cardiac arrest are less likely to survive if the rhythm is asystole or pulseless electrical activity and if the arrest is unwitnessed.
- There are well defined techniques and skills for clinicians to be able to optimally deliver bad news to the loved ones of a patient who experiences a cardiac arrest.
- Indexes of pre-arrest clinical factors can provide valuable prognostic information in regards to which patients are unlikely to survive after a cardiac arrest.

M.G. Dickinson, MD (✉)
Richard DeVos Heart & Lung Transplant Program, Frederik Meijer Heart and Vascular Institute, Spectrum Health, 330 Barclay NE, Suite 200, Grand Rapids, MI 49503, USA

Michigan State University, Grand Rapids, MI, USA
e-mail: michael.dickinson@spectrumhealth.org

C.M. Meeusen, MD
Internal Medicine Department, Spectrum Health Butterworth, Grand Rapids, MI, USA

D.L. Maison, MD
Palliative Care, Spectrum Health Medical Group, Spectrum Health, Grand Rapids, MI, USA

S.J. Goodlin, M.W. Rich (eds.), *End-of-Life Care in Cardiovascular Disease*,
DOI 10.1007/978-1-4471-6521-7_7

- Predictors of neurological outcome after resuscitation from a cardiac arrest should involve daily neurological examination supplemented by data from electroencephalograms (EEGs), somatosensory evoked potentials (SSEPs) and neuron specific enolase (NSE) levels. In most cases, outcome cannot be determined until at least 72 h after rewarming from therapeutic hypothermia.

Introduction

Every healthcare professional with hospital experience knows the drama and the drill around a cardiac arrest. It is something that is repeated daily at hospitals all around the world, occurring more than 200,000 times per year in the United States [1]. When a cardiac arrest is declared, the guidelines are clear and provide specific recommendations to ensure that all know their role and what to do [2]. The medical aspects of the cardiac arrest are, however, only one component. Bigger and harder questions and gaps in knowledge and skillsets exist. When should resuscitation attempts be stopped? After the arrest, what should we do? What are the odds that the patient will recover and have a good quality of life post-arrest? What do we tell the families and loved ones shocked by an unexpected cardiac arrest? How do we communicate to them? How should we advise them? What evidence is available to guide us in not just "running the code" but in the greater task of managing all the aspects of care and communication surrounding a cardiac arrest?

Issues Related to the Resuscitation Attempt

When to Stop

The goal of a resuscitation attempt is the return of spontaneous circulation (ROSC). How long should resuscitation be attempted? In the out of hospital setting, clinical prediction rules have been developed to direct when survival is not possible [3]. One such rule showed no survival for:

1. Patients whose arrest was not witnessed by Emergency Medical Services (EMS) personnel,
2. For whom no shock was given, and
3. For whom there was no ROSC after a reasonable attempt at resuscitation.

Such straightforward rules have been harder to establish, however, in the setting of "in-hospital" cardiac arrest. In one report there was no survival to hospital discharge for patients [4]:

1. With asystole or pulseless electrical activity (PEA) as their initial rhythm,
2. In whom the arrest was not witnessed, and
3. For whom 10 min of resuscitation attempt did not result in ROSC.

On the other hand, hospitals that "try harder" may have better survival. In one large observational study, hospitals in which the median duration of resuscitation attempts was in the highest quartile also had higher survival [5]. Furthermore, hospitals with longer resuscitation attempts not only had higher rates of ROSC but also had higher rates of survival to discharge without increased rates of neurologic injury. This suggests that centers that try harder are more successful in their resuscitation efforts. The data also suggest that time alone cannot be used to determine when to terminate resuscitation attempts.

The implication of these findings is that clinical discretion is required rather than specific rules. If the rhythm is asystole or PEA and the arrest is unwitnessed, survival to discharge is less likely. Duration of resuscitation per se should not be used as a criterion; rather failure of robust efforts at resuscitation should be the primary indicator. Later in this chapter, scoring systems to determine prognosis after an arrest will be reviewed.

Technological Advances: Therapeutic Hypothermia and Extracorporeal CPR (ECMO-CPR)

Recent technological advances have added complexities to deciding on appropriate care in patients who fail to have ROSC or who gain ROSC only after prolonged resuscitation attempts. Mild therapeutic hypothermia has been shown to significantly improve the likelihood of good neurological outcome and survival after cardiac arrest [6, 7], and this has created a marked shift in expectations with regard to neurological outcomes after cardiac arrest. Some patients for whom there would traditionally have been thought to be no hope have recovered with good neurological outcome and have returned to functional lives. However, despite the benefits of therapeutic hypothermia, available data indicate that the technology is under-utilized [8].

The growing availability of urgent mechanical circulatory support such as extra-corporeal membrane oxygenation (ECMO) has created the concept of extra-corporeal cardiopulmonary resuscitation (E-CPR) [9, 10]. In this scenario, patients who do not respond to initial efforts at standard advanced cardiac life support are placed urgently on an external heart pump with or without an oxygenator. This can result in meaningful survival with some patients being resuscitated even after long periods of CPR. Studies have shown an overall 29 % survival to hospital discharge. Even among patients with an extreme duration of CPR (>60 min), the survival rate was 18 %. Of those who survived, the vast majority (93 %) had good neurological function.

Although E-CPR technologies are effective, they require coordinated teams that are able to rapidly deploy them on short notice. Further, survival rates are only around 30 %, and while ideal patient selection criteria have not yet been defined, discretion is advised in the use of these technologies, which require large amounts of equipment and staff resources. Therefore, use of this E-CPR should be applied to situations in which high quality chest compressions have been continuously administered to patients who have a reasonable likelihood of surviving the massive physiologic insult that occurs after a prolonged cardiac arrest.

In summary, patients with an unwitnessed arrest and an initial rhythm of PEA or asystole who fail to regain ROSC despite resuscitation attempts are unlikely to survive. Almost all patients who do not follow verbal commands after ROSC should undergo therapeutic hypothermia with the goal of minimizing anoxic brain injury [2]. In hospitals with E-CPR capability, implementation of this technology should be considered in appropriately selected patients and patients without major comorbid conditions even in the absence of ROSC.

When Resuscitation Fails: How to Deliver Bad News

How to Communicate the News of a Patient's Death

When resuscitation fails after an unexpected cardiac arrest, it is the responsibility of the physician to deliver the bad news to the patient's family and loved ones. Physician training does not generally prepare physicians for executing this responsibility. While it is always difficult to convey news of a sudden death, there is evidence that a structured program can help reduce the pain experienced by families and friends [11]. Preparing survivors for the possibility of death in stages during the course of resuscitation can be helpful. It is also important to survivors to perceive that the medical staff shares in the grief over the death of their loved one [12]. While staff does not have to become emotional, they should convey a sense of grief that the resuscitation was ultimately not successful. Table 7.1 describes key steps in a program to facilitate the delivery of bad news.

How to Communicate the News of a Non-fatal Cardiac Arrest

If the patient has survived, communication of the event to the patient or their loved ones can also be challenging. Techniques outlined in Table 7.1 may still apply. Additionally, research suggests that the Ask-Tell-Ask approach is an effective way to communicate difficult information [13]. In this method, the patient or family members are asked what they understand (Ask), important information or details are communicated in non-medical terms (Tell), and the provider then asks participants to repeat what they have

Table 7.1 Protocol to deliver news of sudden death

1. Social work (SW) or support person calls family (friends/loved ones) to the hospital. Do not deliver news of death over the phone unless impractical to do otherwise
2. Family placed in a quiet private room (appropriate setting)
3. SW begins to prepare family by stressing the gravity of the patient's condition and the poor prospects for survival
4. Physician prepares to deliver the bad news by reviewing the facts of the case before entering the room (have accurate information, be able to answer questions)
5. Physician delivers the news in an unambiguous manner. Do not avoid words such as "dead" or "died"
6. Have sympathy in communication. Deliver the news in a manner that conveys "someone cares that their loved one died"
7. Provide reassurance that the deceased did not suffer and that everything that could be done was done
8. Have resources and information to help with grief and logistics of the "next steps"

been told to ensure that the information was correctly received (Ask) [14]. This approach has the advantage of hearing and then addressing the families' deepest concerns or fears first. In so doing, it may be possible to relieve their anxiety so that they can hear and comprehend additional information that needs to be provided. The second "ask" ensures that all important information was conveyed clearly.

After the cardiac arrest, frequent communication with family and friends of a resuscitated patient is important for setting expectations. In the era of therapeutic hypothermia, post-resuscitation care now extends through a period of 3–7 days. This period should be viewed as a valuable therapeutic window for the delivery of grief counseling and support to family and friends. Palliative care teams including social work, pastoral care and palliative care specialists can assist in guiding family and friends through the stages of grief. Team members skilled at compassionate communication should accurately convey "realistic expectations" but "not deprived of hope" when uncertainty exists. If prognostic indicators become poor, the family and friends can be guided through a "weaning of hope" until acceptance of the loss of their loved one can occur.

Issues After Resuscitation

Prognostication to Assist in Potential End of Life Decision Making

Which patients will survive to discharge after an in-hospital cardiac arrest?

Data indicate that physicians are not able to determine who will survive after cardiac arrest based on clinical judgment alone. In one study, when physicians were given clinical summaries and asked to predict which patients would or would not survive, their predictions were no better than chance [15].

Factors that have been associated with survival include [16]:

- Occurrence of cardiac arrest within 24 h of admission to the hospital
- Short duration of CPR
- Absence of pre-arrest:
  - Cardiogenic shock
  - Sepsis
  - Acute renal failure
  - Cancer
  - Pneumonia

Very poor survival has been noted in patients with both age over 60 years and CPR efforts lasting more than 10 min [17]. In a comprehensive meta-analysis, pre-arrest factors predictive of poor survival included older age, inability to perform activities of daily living (ADLs) prior to hospitalization, abnormal mental status, abnormal renal function, hypotension, and malignancy [18]. The likelihood of survival declines progressively with age, and there is a prominent drop off in survival after the age of 75 years.

Three scores have been developed to predict survival following in-hospital cardiac arrest. The Pre-Arrest Morbidity (PAM) Score was derived from a prospective series of in-hospital cardiac arrests at a teaching hospital in the United States [19]. The investigators found that pre-arrest hypotension, renal dysfunction and age > 65 were strong univariate predictors of poor outcome. Other factors included in the PAM score are listed in Table 7.2. Patients with a PAM score of 7 or greater had a

Table 7.2 Pre-arrest morbidity score*

Clinical characteristic	Point value
Hypotension	3
Azotemia (BUN > 50 or Cr >2.5 mg/dl)	3
Malignancy	3
Pneumonia	3
Homebound lifestyle	3
Angina pectoris	1
Acute myocardial infarction	1
Heart failure (NYHA III, IV)	1
S3 gallop	1
Oliguria (<300 ml/day)	1
Sepsis	1
Mechanical ventilation	1
Recent cerebrovascular event	1
Coma	1
Cirrhosis	1

From George et al. [19]. Reprinted with permission from Elsevier Limited

*See text for details on how to interpret the score

BUN blood urea nitrogen, *Cr* serum creatinine, *NYHA* New York Heart Association very low likelihood of long term survival (<15 %), and no patient with a score of 8 or higher survived. For example, a patient with a serum creatinine >2.5 mg/dl, homebound lifestyle, NYHA class III heart failure, and an S3 gallop at admission would not be expected to survive an in-hospital cardiac arrest. The PAM score has been validated in three different patient populations and in each of these studies, no patients with scores >8 survived [19–21].

The modified PAM index (MPI) added dementia (2 points), reduced cancer from 3 to 2 points, assigned 1 point for myocardial infarction (MI) only if the arrest occurred more than 48 h later, and removed cirrhosis from the score [22]. While no direct validation of the PAM vs. the MPI was performed, the authors based the modifications on a robust data pool derived from 32 studies.

The Prognosis After Resuscitation (PAR) score was developed from a meta-analysis of 14 studies of survival following in hospital post cardiac arrest and assigns a score based on 8 variables, each of which is assigned a score ranging from −2 to 10 points as indicated below [23].

- Metastatic cancer: 10
- Sepsis: 5
- Dependent functional status: 5
- Non-metastatic cancer: 3
- Pneumonia: 3
- Serum creatinine ≥ 1.5 mg/dl: 3
- Age >70 years: 2
- Acute MI: −2

When the PAR score was applied to a cohort of 218 patients, 37 (20.1 %) had a score greater than 8 and none of these patients survived.

The utility of the PAM, MPI, and PAR scores has been assessed in independent populations. In one series from a hospital in England, each score was able to reliably predict non-survivors with 100 % specificity [24]. However, the sensitivity was low and varied from 20 to 30 %. Thus, the majority of patients who do not survive are not reliably identified by the scores. The authors noted that if either the PAR or MPI score was greater than 6, the sensitivity for predicting death increased to 41 % with no loss of specificity (i.e., all patients with a score >6 died). However, this has not been independently validated. Since each of the scores selected slightly different populations, use of all 3 scores could maximize identification of patients for whom resuscitation attempts are likely to be futile.

Another large meta-analysis evaluated these scores along with the APACHE-II (Acute Physiology and Chronic Health Evaluation) score [25] and found similar results; i.e., there was high specificity at the expense of poor sensitivity [18]. Patients with a PAM >8, a PAR >8, or an MPI >6 were not likely to survive. APACHE-II had less robust specificity, and an APACHE-II score >20 was associated with a 4.8 % chance of survival.

Predicting Neurological Outcome After In-Hospital Cardiac Arrest

Prediction of a successful neurological outcome after ROSC is challenging in the era of therapeutic hypothermia. As with physicians' ability to predict cardiac arrest survival, neurologists' clinical judgment for predicting neurological outcomes lacks reliability. One case series demonstrated that in several patients for whom board certified neurologists had predicted grave prognoses, full recovery was achieved [26]. Many of the standard rules to predict adverse prognosis do not apply in the era of therapeutic hypothermia. For example, the absence of an extensor response to pain on day 3 after cardiac arrest is considered a grave finding [27]. In patients treated with therapeutic hypothermia, however, the extensor pain response was absent in 10 % of patients with satisfactory neurological recovery, suggesting that this sign alone cannot be used to decide when to withdraw support [28]. Generalized myoclonus on the first day after a cardiac arrest has also been strongly predictive of non-recovery of neurological function [29]. There are now, however, reports of patients treated with therapeutic hypothermia who survived with good neurological outcome despite early generalized myoclonus [30]. Neurological examination at 72 h after cardiac arrest has often been the standard to determine if neurological function will return. In patients receiving therapeutic hypothermia, this is no longer considered reliable [31–33]. In one series, 6 of 34 patients with persistent coma 4–5 days after arrest later regained consciousness and were alive 6 months later. Neuron-specific enolase (NSE) is a serologic biomarker and a value >33 μmol/L has been considered a reliable predictor of poor outcome [34]. However, in a more recent prospective trial of patients treated with hypothermia, 10 of 99 patients with an NSE level >33 μmol/L had a good neurological outcome [28].

To provide insight into how best to assess neurological prognosis in patients receiving therapeutic hypothermia, Friberg et al. summarized the available data and suggested a multimodality approach with continuous evaluation including daily neurological examinations and simplified electroencephalographic (EEG) recordings [33]. Continuous amplitude integrated EEGs (aEEGs) can provide prognostic information, especially in the presence of 2 distinct patterns [35]. A continuous pattern on aEEG is defined as continuous cortical activity within the delta, theta, and/or alpha bands of the patient's standard EEG. A suppression burst pattern on aEEG is seen as high voltage bursts (>50 μV) of slow waves interrupted by suppression (low amplitude <10 μV lasting >1 s). In patients whose aEEGs showed a continuous pattern, 90 % regained consciousness. Conversely, patients with a suppression burst pattern did not survive. EEGs are also important for detecting electrographic status epilepticus (ESE) [36]. Somatosensory Evoked Potentials (SSEP) can also be helpful. The N20 potential is an electrical signal measured over the somatosensory cortex during contralateral wrist stimulation, thus suggesting intact cortical sensory activity. Bilateral loss of N20 potentials is associated with very poor neurological outcome in cardiac arrest patients who underwent treatment with hypothermia and rewarming [27]. NSE levels measured over time can provide additional information, especially if they remain low [33]. If NSE levels are consistently <33 μmol/L, an investigation for other causes of persistent coma, such as ESE or prolonged sedation effect, is advised. Computed tomographic (CT) imaging is valuable for detecting massive cerebral edema and herniation. Quantitative CT within the first 24 h after arrest to measure the gray to white matter attenuation ratio (GWR) can be helpful. In one study, only 2 of 58 patients (3.4 %) with a GWR <1.2 survived, and no patients (0 of 20) with a GWR <1.1 survived [37]. Friberg et al. suggested that withdrawal of care could appropriately be performed under the following circumstances:

1. Brain death from cerebral herniation.
2. Generalized myoclonus (face and extremities) in the first 24 h AND bilateral absence of N20 peaks on median nerve somatosensory evoked potentials (SSEP) after rewarming.
3. At 72 h after warming, persistent coma with Glasgow Motor Scale 1–2 AND one of the following:

   (a) Absent N20 peaks on SSEP (as above).
   (b) Treatment refractory status epilepticus.
   (c) No improvement in neurological status for an additional 1–2 days.

The Best Treatment Is Prevention

While cardiac arrests are an inevitable occurrence in most hospitals, in many cases in-hospital cardiac arrest reflects a failure of the healthcare system on multiple levels. First, cardiac arrest is often the terminal event for patients for whom death is imminent. In an ideal healthcare system, such patients would be identified prospectively, compassionate and honest discussions would be held, and “do not resuscitate” orders would be placed. Ebell et al. suggest that prognostic indicators such as the PAR score be routinely calculated on all patients [18]. Using this information, clinicians could then have an informed code status discussion with patients and families. Based on the score, patients could be classified into one of three categories: (1) Futility (survival to discharge after cardiac arrest is so unlikely that most would consider it futile to attempt resuscitation), (2) Uncertain (survival is unlikely but some may choose resuscitation based on personal values and goals), and (3) Beneficial (survival to discharge is of average or greater likelihood). When clinicians discuss code status with their patients, providing information on the probability of survival with good neurological function could greatly facilitate informed decision-making about end of life choices. This in turn could reduce the number of inevitably unsuccessful resuscitation attempts.

Second, an in-hospital cardiac arrest may represent a “failure to rescue” (FTR). A study at a United States academic hospital found that 14 % of in-hospital cardiac arrests were iatrogenic [38]. Of these, over half were estimated to have been preventable by closer attention to details of the patient’s history, examination or laboratory findings. Applying this analysis to the estimated 200,000 in-hospital cardiac arrests that occur annually in the United States, there are up to 28,000 iatrogenic cardiac arrests and 14,000 fully preventable cardiac arrests per year.

A third scenario, which is even more frequent than iatrogenic cardiac arrest, is that a decline in a patient's status occurred that presaged impending cardiac arrest. In some cases, these clinical changes might have been detected and intervened upon to prevent culmination in cardiac arrest. Alternatively, further discussions of code status might be warranted in some cases.

Potentially preventable FTR events have been classified into two categories: timely response (promptly recognizing the complication or problem) and appropriate response (a correct and effective treatment is implemented) [39]. Rapid response teams may be associated with prevention of some in-hospital cardiac arrests [40]. Ideally, in-hospital cardiac arrests should be systematically and routinely reviewed to identify FTR with the goal of preventing similar events in other patients. Such a review should include the following elements:

1. Should this patient have had CPR? Was there a missed opportunity to discuss end of life preferences and goals of care prior to the event? Was this patient one in whom non-survival was inevitable based on risk scores?
2. Were there markers of physiologic distress that could have been detected in the minutes to hours prior to the arrest? Such indicators might include tachypnea, tachycardia, hypotension, hypoxemia, oliguria, delirium, acidemia, alkalosis, or other vital sign or laboratory parameters.
3. Were there opportunities for a more robust response to abnormal indicators or concerns prior to the arrest? Were nursing concerns heeded and responded to in the most effective way possible? What alternative responses might have prevented the arrest?
4. How well did the system perform in communicating the details of the arrest and in guiding the family through post-arrest care?

Conclusions

There is a substantial evidence base to guide clinicians in optimizing care for the hospitalized patient who experiences an unexpected cardiac arrest. While there are no firm rules about when to terminate efforts, resuscitation attempts are less likely to be successful in patients with non-shockable rhythms and unwitnessed arrests. Therapeutic hypothermia and E-CPR are new technologies that can improve the likelihood of long term survival with good neurological outcome. Techniques and skills have been developed to assist clinicians with the difficult task of sharing bad news with families after an arrest. Prognostic tools are available using both pre-arrest clinical indicators and post-arrest tests of neurological function to help guide decision-making after resuscitation. Finally, clinicians should use available information to guide systems of care to ensure that when a cardiac arrest occurs, resuscitation efforts are indicated and that pre-arrest warning signs were identified and appropriately managed with the goal of preventing as many arrests as possible.

References

1. Go AS, Mozaffarian D, Roger VL, Benjamin EJ, Berry JD, Borden WB, Bravata DM, Dai S, Ford ES, Fox CS, Franco S, Fullerton HJ, Gillespie C, Hailpern SM, Heit JA, Howard VJ, Huffman MD, Kissela BM, Kittner SJ, Lackland DT, Lichtman JH, Lisabeth LD, Magid D, Marcus GM, Marelli A, Matchar DB, McGuire DK, Mohler ER, Moy CS, Mussolino ME, Nichol G, Paynter NP, Schreiner PJ, Sorlie PD, Stein J, Turan TN, Virani SS, Wong ND, Woo D, Turner MB. Executive summary: heart disease and stroke statistics–2013 update: a report from the American Heart Association. Circulation. 2013;127:143–52.
2. Field JM, Hazinski MF, Sayre MR, Chameides L, Schexnayder SM, Hemphill R, Samson RA, Kattwinkel J, Berg RA, Bhanji F, Cave DM, Jauch EC, Kudenchuk PJ, Neumar RW, Peberdy MA, Perlman JM, Sinz E, Travers AH, Berg MD, Billi JE, Eigel B, Hickey RW, Kleinman ME, Link MS, Morrison LJ, O'Connor RE, Shuster M, Callaway CW, Cucchiara B, Ferguson JD, Rea TD, Vanden Hoek TL. Part 1: executive summary: 2010 American Heart Association Guidelines for Cardiopulmonary Resuscitation and Emergency Cardiovascular Care. Circulation. 2010;122:S640–56.
3. Morrison LJ, Verbeek PR, Zhan C, Kiss A, Allan KS. Validation of a universal prehospital termination of resuscitation clinical prediction rule for advanced and basic life support providers. Resuscitation. 2009;80:324–8.
4. van Walraven C, Forster AJ, Stiell IG. Derivation of a clinical decision rule for the discontinuation of in-hospital cardiac arrest resuscitations. Arch Intern Med. 1999;159:129–34.
5. Goldberger ZD, Chan PS, Berg RA, Kronick SL, Cooke CR, Lu M, Banerjee M, Hayward RA, Krumholz HM, Nallamothu BK. Duration of resuscitation efforts and survival after in-hospital cardiac arrest: an observational study. Lancet. 2012;380:1473–81.
6. The hypothermia after Cardiac Arrest Study Group. Mild therapeutic hypothermia to improve the neurologic outcome after cardiac arrest. N Engl J Med. 2002;346: 549–56.
7. Bernard SA, Gray TW, Buist MD, Jones BM, Silvester W, Gutteridge G, Smith K. Treatment of comatose survivors of out-of-hospital cardiac arrest with induced hypothermia. N Engl J Med. 2002;346:557–63.
8. Mikkelsen ME, Christie JD, Abella BS, Kerlin MP, Fuchs BD, Schweickert WD, Berg RA, Mosesso VN, Shofer FS, Gaieski DF. Use of therapeutic hypothermia after in-hospital cardiac arrest*. Crit Care Med. 2013;41:1385–95.
9. Bagai J, Webb D, Kasasbeh E, Crenshaw M, Salloum J, Chen J, Zhao D. Efficacy and safety of percutaneous life support during high-risk percutaneous coronary intervention, refractory cardiogenic shock and in-laboratory cardiopulmonary arrest. J Invasive Cardiol. 2011;23:141–7.
10. Chen YS, Lin JW, Yu HY, Ko WJ, Jerng JS, Chang WT, Chen WJ, Huang SC, Chi NH, Wang CH, Chen LC, Tsai PR, Wang SS, Hwang JJ, Lin FY. Cardiopulmonary resuscitation with assisted extracorporeal life-support versus conventional cardiopulmonary resuscitation in adults with in-hospital cardiac arrest: an observational study and propensity analysis. Lancet. 2008;372:554–61.
11. Adamowski K, Dickinson G, Weitzman B, Roessler C, Carter-Snell C. Sudden unexpected death in the emergency department: caring for the survivors. CMAJ. 1993;149:1445–51.
12. Jones WH, Buttery M. Sudden death: survivors' perceptions of their emergency department experience. J Emerg Nurs. 1981;7:14–7.
13. Back AL, Arnold RM, Baile WF, Tulsky JA, Fryer-Edwards K. Approaching difficult communication tasks in oncology. CA Cancer J Clin. 2005;55:164–77.
14. Kemp EC, Floyd MR, McCord-Duncan E, Lang F. Patients prefer the method of "tell back-collaborative inquiry" to assess understanding of medical information. J Am Board Fam Med. 2008;21:24–30.
15. Ebell MH, Bergus GR, Warbasse L, Bloomer R. The inability of physicians to predict the outcome of in-hospital resuscitation. J Gen Intern Med. 1996;11:16–22.

16. Rozenbaum EA, Shenkman L. Predicting outcome of inhospital cardiopulmonary resuscitation. Crit Care Med. 1988;16:583–6.
17. Schultz SC, Cullinane DC, Pasquale MD, Magnant C, Evans SR. Predicting in-hospital mortality during cardiopulmonary resuscitation. Resuscitation. 1996;33:13–7.
18. Ebell MH, Afonso AM. Pre-arrest predictors of failure to survive after in-hospital cardiopulmonary resuscitation: a meta-analysis. Fam Pract. 2011;28:505–15.
19. George ALJ, Folk BP, Crecelius PL, Campbell WB. Pre-arrest morbidity and other correlates of survival after in-hospital cardiopulmonary arrest. Am J Med. 1989;87:28–34.
20. Cohn EB, Lefevre F, Yarnold PR, Arron MJ, Martin GJ. Predicting survival from in-hospital CPR: meta-analysis and validation of a prediction model. J Gen Intern Med. 1993;8:347–53.
21. O'Keeffe S, Ebell MH. Prediction of failure to survive following in-hospital cardiopulmonary resuscitation: comparison of two predictive instruments. Resuscitation. 1994;28:21–5.
22. Dautzenberg PL, Broekman TC, Hooyer C, Schonwetter RS, Duursma SA. Review: patient-related predictors of cardiopulmonary resuscitation of hospitalized patients. Age Ageing. 1993;22:464–75.
23. Ebell MH. Prearrest predictors of survival following in-hospital cardiopulmonary resuscitation: a meta-analysis. J Fam Pract. 1992;34:551–8.
24. Bowker L, Stewart K. Predicting unsuccessful cardiopulmonary resuscitation (CPR): a comparison of three morbidity scores. Resuscitation. 1999;40:89–95.
25. Knaus WA, Draper EA, Wagner DP, Zimmerman JE. APACHE II: a severity of disease classification system. Crit Care Med. 1985;13:818–29.
26. Yannopoulos D, Kotsifas K, Aufderheide TP, Lurie KG. Cardiac arrest, mild therapeutic hypothermia, and unanticipated cerebral recovery. Neurologist. 2007;13:369–75.
27. Zandbergen EG, de Haan RJ, Stoutenbeek CP, Koelman JH, Hijdra A. Systematic review of early prediction of poor outcome in anoxic-ischaemic coma. Lancet. 1998;352:1808–12.
28. Bouwes A, Binnekade JM, Kuiper MA, Bosch FH, Zandstra DF, Toornvliet AC, Biemond HS, Kors BM, Koelman JH, Verbeek MM, Weinstein HC, Hijdra A, Horn J. Prognosis of coma after therapeutic hypothermia: a prospective cohort study. Ann Neurol. 2012;71:206–12.
29. Thomke F, Marx JJ, Sauer O, Hundsberger T, Hagele S, Wiechelt J, Weilemann SL. Observations on comatose survivors of cardiopulmonary resuscitation with generalized myoclonus. BMC Neurol. 2005;5:14.
30. Lucas JM, Cocchi MN, Salciccioli J, Stanbridge JA, Geocadin RG, Herman ST, Donnino MW. Neurologic recovery after therapeutic hypothermia in patients with post-cardiac arrest myoclonus. Resuscitation. 2012;83:265–9.
31. Blondin NA, Greer DM. Neurologic prognosis in cardiac arrest patients treated with therapeutic hypothermia. Neurologist. 2011;17:241–8.
32. Cronberg T, Rundgren M, Westhall E, Englund E, Siemund R, Rosen I, Widner H, Friberg H. Neuron-specific enolase correlates with other prognostic markers after cardiac arrest. Neurology. 2011;77:623–30.
33. Friberg H, Rundgren M, Westhall E, Nielsen N, Cronberg T. Continuous evaluation of neurological prognosis after cardiac arrest. Acta Anaesthesiol Scand. 2013;57:6–15.
34. Wijdicks EF, Hijdra A, Young GB, Bassetti CL, Wiebe S. Practice parameter: prediction of outcome in comatose survivors after cardiopulmonary resuscitation (an evidence-based review): report of the Quality Standards Subcommittee of the American Academy of Neurology. Neurology. 2006;67:203–10.
35. Rundgren M, Westhall E, Cronberg T, Rosen I, Friberg H. Continuous amplitude-integrated electroencephalogram predicts outcome in hypothermia-treated cardiac arrest patients. Crit Care Med. 2010;38:1838–44.
36. Rittenberger JC, Popescu A, Brenner RP, Guyette FX, Callaway CW. Frequency and timing of nonconvulsive status epilepticus in comatose post-cardiac arrest subjects treated with hypothermia. Neurocrit Care. 2012;16:114–22.

37. Metter RB, Rittenberger JC, Guyette FX, Callaway CW. Association between a quantitative CT scan measure of brain edema and outcome after cardiac arrest. Resuscitation. 2011;82:1180–5.
38. Bedell SE, Deitz DC, Leeman D, Delbanco TL. Incidence and characteristics of preventable iatrogenic cardiac arrests. JAMA. 1991;265:2815–20.
39. Ghaferi AA, Birkmeyer JD, Dimick JB. Variation in hospital mortality associated with inpatient surgery. N Engl J Med. 2009;361:1368–75.
40. Van Voorhis KT, Willis TS. Implementing a pediatric rapid response system to improve quality and patient safety. Pediatr Clin North Am. 2009;56:919–33.

Chapter 8
End-of-Life Care in Skilled Nursing Facilities

Corrine Y. Jurgens and Diane K. Pastor

Abstract End-of-life care for patients with cardiovascular illness living in skilled nursing facilities (SNF) is associated with challenges related to prognosis and symptom management. This chapter provides an overview of the components of end-of-life care for patients with cardiovascular disease. It includes advanced care planning, shared decision making, symptom management, and bereavement appropriate for this complex population. Among approximately two million SNF residents, cardiovascular disease is the largest diagnostic category and heart failure is common. Symptoms of heart failure are burdensome and negatively affect quality of life. Symptom management with maximized medical guideline-based therapy as tolerated is important from time of diagnosis to end-of-life. As determining prognosis among patients with heart failure is difficult, using uncontrollable symptoms as a trigger for end-of-life care may be prudent. An interdisciplinary approach using the skills of nursing, medicine, social work, chaplains and volunteers is recommended for resident and family satisfaction with end-of-life. End-of-life care focuses on needs of the resident and family by encouraging and supporting shared decision-making.

Keywords Heart failure • Skilled nursing facilities • Palliative care • Symptom management • Older adults • End-of-life care • Hospice • Advanced directives • Shared decision-making • Grief • Bereavement

C.Y. Jurgens, PhD, RN, ANP-BC, FAHA (✉)
School of Nursing, Stony Brook University,
HSC L2-246, Stony Brook, NY 11794-8240, USA
e-mail: corrine.jurgens@stonybrook.edu

D.K. Pastor, PhD, MBA, NP-C
Adult Health Program, School of Nursing, Stony Brook University, Stony Brook, NY, USA

S.J. Goodlin, M.W. Rich (eds.), *End-of-Life Care in Cardiovascular Disease*,
DOI 10.1007/978-1-4471-6521-7_8

Key Points

- One in four adults will experience end-of life in long-term care
- Advanced care planning for residents with cardiovascular disease includes contingency planning for device management, hospitalization, and resuscitation
- Prognostication of impending death is challenging. Uncontrollable symptoms may be useful in transitioning to hospice care
- Symptom assessment and management are central to achieving a good death

As our population ages, long-term care facilities (nursing homes, skilled nursing facilities (SNF)) are increasingly the setting for end-of-life. With 25 % of older Americans dying in SNF [1, 2], quality end-of-life care is an integral component of care in accordance with resident and family preferences. The terms palliative care and hospice care often are used interchangeably to describe comfort care at end-of-life. For the purpose of this chapter, nonhospice palliative care refers to interventions to support quality of life over months to years for those with life-limiting cardiovascular illness. Hospice care refers to care when death is likely to occur in 6 months or less [3–5].

SNF residents are among the oldest old and cognitive impairment and multiple comorbid illnesses are common in this population [6]. Among approximately two million SNF residents, cardiovascular disease is the largest diagnostic category and heart failure is common [7, 8].

Heart failure is a chronic and progressive syndrome where structural or functional impairment of the heart affects ventricular filling or ejection of blood [9]. In addition to advanced age, several comorbid illnesses common in older adults such as ischemic heart disease, hypertension and valvular disease increase risk for developing heart failure (Fig. 8.1) [10, 11]. Symptoms of heart failure are burdensome and negatively affect quality of life [12–15]. To support quality of life, symptom management (non-hospice palliative care) with maximized medical guideline-based therapy is important from time of diagnosis to end-of-life [3]. Timely transition to hospice care is based on prognosis, symptoms and patient and family preferences.

Prognostic Models

A first step in providing quality end-of-life care is identification of prognosis for residents in SNF [16]. However, heart failure is frequently associated with an unpredictable trajectory of exacerbations, plateaus or periods of stability, and functional decline [17]. Compared with other terminal illnesses such as cancer where time to end-of-life is predicted in months, predicting end-of-life for patients with heart failure may be measured in years. As a result, heart failure presents prognostic

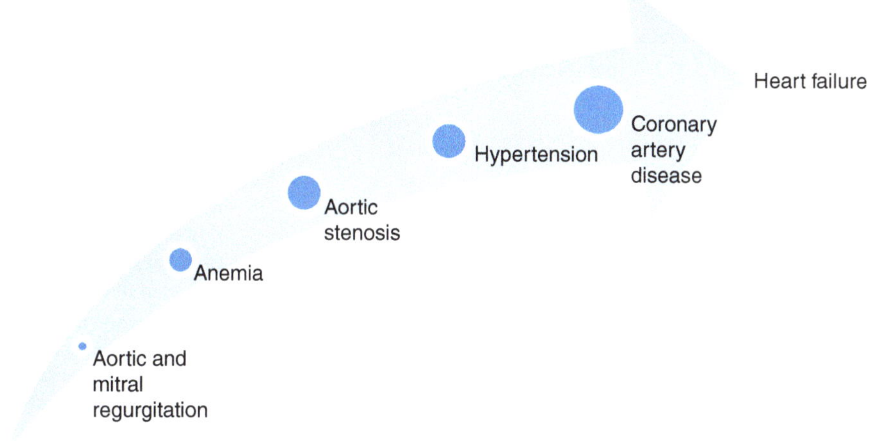

Fig. 8.1 Pathways leading to heart failure in older adults

challenges to the SNF health care team. Understanding the variable nature of the course of heart failure is necessary to appreciate the difficulty in applying a palliative care model to care of heart failure patients. Several prognostic models using both single and multiple variables have been studied, but issues with over or under estimation of mortality risk have been reported (Table 8.1) [3, 18, 19].

The Seattle Heart Failure Model includes the greatest number of variables and was validated in large outpatient populations. Levy and colleagues developed and validated the Seattle model to predict 1-, 2-, and 3-year survival in heart failure patients with the use of clinical status, therapy and laboratory parameters [20]. The model allows estimation of the benefit of adding medications or devices to an individual patient's therapeutic regimen. An online calculator is available at http://www.seattleheartfailuremodel.org. The model may not be generalizable to patients with dementia which is common in the SNF population. Similarly, other major comorbid illnesses such as cancer or renal failure are not accounted for in the model [20].

The Gold Standard Framework was developed to identify people nearing the end-of-life to facilitate appropriate palliative and end-of-life care. The model has three components beginning with the surprise question: "Would you be surprised if the patient were to die in the next few months, weeks or days" [21]. General indicators of decline include decreasing activity tolerance and increasing symptom burden. Heart failure specific indicators are the third component which includes poor functional status, repeated admissions, and high symptom burden despite optimized therapies.

In a comparison of the Seattle and Gold Standard models, neither was found to be accurate in predicting mortality at 12 months [18]. However, renal function measured using serum creatinine was the most sensitive and specific factor for

Table 8.1 Heart failure prognostic models

Model	Prediction	Features	Considerations
Seattle Heart Failure Model [20]	1, 2, 3 year survival	Uses clinical variables, therapeutic interventions, and devices:	May not apply to SNF populations (e.g. patients with dementia)
		1. Age	Developed in patients with predominantly left ventricular systolic HF
		2. Sex	Does not include comorbid illnesses
		3. Weight 4. Ejection fraction 5. NYHA class 6. Systolic blood pressure 7. Daily diuretic dose 8. Serum sodium 9. Hemoglobin 10. % lymphocytes 11. Uric acid 12. Total cholesterol 13. Medications (e.g. angiotensin receptor blockers, angiotensin receptor blockers, β-blockers, statins, aldosterone blockers) 14. Ischemic etiology 15. Devices Graphic impact of adding or removing interventions Validated with large outpatient populations	Renal function not included in the model as it was tested in patients without significant renal dysfunction
Gold Standard Framework [21]	Indicator for nearing end-of-life and need for palliative/hospice care	Uses three components to predict declining HF status: 1. Decreasing activity tolerance 2. Increasing symptom burden 3. HF specific indicators include: (a) NYHA III/IV (b) admissions to inpatient settings (c) Level of symptom burden	May overestimate number of patients nearing end-of-life [18]

predicting death at 12 months. Lower serum creatinine predicted lower mortality at 1 year (<140 μmol/l).

An accurate prognostic model specifically for use in the SNF population is currently not known. Furthermore, some models incorporate variables that may not be readily available in SNF. As difficulty in prognosticating mortality delays initiation of palliative and hospice care, it may be more prudent to use uncontrollable symptoms as a trigger for end-of-life care [22, 23].

End-of-Life Care Components

To provide quality end-of-life care for HF patients in SNF, programs that address the goals of this care for residents and their care giving families must be developed. Components of high quality end-of-life care programs include:

- Focusing on needs of the resident and family
- Encouraging and supporting shared decision-making
- Documenting advanced directives to support quality of life and a good death
- A pain-free and aware death and acknowledgement of the imminence of death [24]
- A framework acknowledging cultural implications about illness and dying and providing for seamless transitions of care
- A holistic approach to meet SNF residents' and their families' physical, psychological/emotional, spiritual needs and preferences [25, 26]

Shared Decision-Making

Effective communication and shared decision-making among the health care team, residents and families is integral to achieving quality end-of-life care for residents with life-limiting cardiovascular disease including heart failure. Shared decision-making differs from patient centered care in which communication focuses on the needs and emotions of the patient [27]. In shared decision-making, the resident (and families) and provider(s) collaborate to make treatment decisions. Information sharing moves in both directions with the goal of equal empowerment to make decisions. Participation by the healthcare team (medicine, nursing and other personnel involved in care) may make the process easier for the resident and family [28, 29]. Training of healthcare providers in shared decision-making is important. Lack of skill and confidence in shared decision-making among providers is a barrier to meaningful advance care planning [30, 31]. (See also Chap. 2 about communication and decision-making.)

Shared decision-making preferably continues from admission to SNF care until end-of-life. To enable active participation in decision-making at end-of-life, it is necessary that the resident, family, and healthcare providers recognize and accept that death is impending. For residents and families, knowledge and acceptance that heart failure is a life-limiting illness is not always clear or desired [24, 32]. Although prognostication is challenging, clinical signs such as progressive renal dysfunction, cachexia, escalating diuretic doses, intractable symptoms and declining functional status are potential indicators of impending death [33].

Advance Directives

Advance directives including identification of a healthcare proxy and contingency planning are important considerations at time of admission to a SNF [34]. Do Not Resuscitate (DNR) orders are important but insufficient for outlining resident wishes. Contingency planning specific to heart failure includes preferences related to acute decompensation, hospitalization, implanted cardiac rhythm devices and cardiac arrest [25, 35, 36]. Discussions regarding implantable defibrillators are particularly important as over 20 % of patients receive shocks at end-of-life and shocks are associated with decreased quality of life [35, 37]. In the event a resident does not have a contingency plan or in the absence of readily accessible onsite electrophysiological deactivation equipment, implantable defibrillators can be deactivated by placing a donut magnet over the device. Accordingly, donut magnets should be readily available in SNF. The donut magnet should be left in place until electrophysiological equipment is available to permanently deactivate the device [35].

Although advance directives including DNR documentation are an important consideration for SNF residents, 66 % were reported not to have DNR or Do Not Hospitalize (DNH) directives in a sample of 1,926,742 Medicare beneficiaries in 2001 [38]. The frequency of directives varies widely by state and race [34, 38]. African-Americans are the least likely to have either a DNR (odds ratio 0.35) or DNH (odds ratio 0.26) directive. Factors associated with an increased likelihood of having either a DNR or DNH advanced directive are outlined in Table 8.2.

Table 8.2 Factors associated with increased likelihood of DNR or DNH directives

Age 95 years and older
Caucasian
Functional dependence
Impaired cognition
Neoplasm
Heart failure
Chronic obstructive lung disease
Dementia
Arrhythmia
Coma
Hypertension

Quality of Life

As heart failure progresses, the focus shifts away from life-extending therapeutics to a focus on supporting and maintaining quality of life [3]. However, knowledge deficits about palliative care exist among patients with heart failure, their caregiving families, and healthcare providers [39]. It is important to explain to residents and families that medical therapies to manage symptoms of heart failure will continue including therapies for symptoms arising from comorbid illnesses. Early initiation of palliative care supports higher satisfaction among residents and their families at end-of-life [40].

Communication patterns used by heart failure patients with their healthcare providers were explored by Green and colleagues in a narrative systematic review of 18 publications [41]. A majority of patients with heart failure wished to have a discussion of both prognostic indicators related to their illness and their feelings about end-of-life issues. Exchange of information between patients and providers may be impeded if the patient has cognitive deficits or fatigue [42]. Therefore, it is prudent to include designated family members or healthcare proxies in these discussions.

Symptom Management

Symptoms of advanced heart failure can be distressing and negatively affect quality of life for patients and their families. The most common symptoms and sources of distress among patients with advanced heart failure are dyspnea, pain, fatigue, anxiety, and depression [13, 17, 25, 43]. In a study of dying residents in long-term care in the last three months of life (n = 674), half had pain and shortness of breath. Difficulty with food and fluid intake was problematic for the majority (75 %) and one-third had dehydration and weight loss [44]. Barriers to symptom management in SNF care include resident factors (e.g. inability to report symptoms) and system factors (e.g. lack of knowledge and or procedures for symptom assessment and management). For residents unable to self-report symptoms, evidence supports that staff and family assessment of quality of care of symptoms are similar [44].

As a first step in symptom management, heart failure medical therapies should be optimized as per published guidelines [9, 45]. However, residents with hypotension or renal failure may not tolerate maximal guideline therapies. The symptom experience encompasses more than simply a physical sensation. Symptoms are generally multiple in numbers and incur physical sensations, cognitive interpretations, and affective reactions to symptoms [46, 47]. It follows that interventions to manage symptoms are therefore multidimensional in nature. (See also Chap. 3.)

Dyspnea

Dyspnea is one of the most common symptoms in advanced heart failure. Interventions for dyspnea are directed at reducing ventilatory demand, ventilatory impedance, and improving central perceptions of dyspnea and inspiratory muscle function [48]. For residents with refractory dyspnea, low dose opiates can be used to reduce ventilatory demand and improve the resident's perception of dyspnea [49, 50]. To assess benefit and observe for adverse effects of opiates, it is appropriate to begin with short-acting opiates. For example, oral morphine starting at 2.5–5 mg every 4 h can be given and slowly titrated upward as needed. Respiratory depression secondary to opiates is generally avoided with slow up-titration [25]. It is appropriate to treat continuous symptoms with long-acting opioids. Among patients with renal dysfunction, metabolites of morphine can accumulate resulting in signs of toxicity such as myoclonus, anxiety and delirium. In such cases, alternate choices for managing dyspnea include hydromorphone or methadone [25].

Although commonly perceived by patients and clinicians to be a useful treatment for dyspnea, oxygen is appropriate only for patients with hypoxia [51, 52]. Nonpharmacological strategies for dyspnea include maintaining a calm environment to prevent sympathetic stimulation, sitting the resident up leaning forward and using a fan directed at the resident's face [53–55]. For residents with cardiomegaly, avoiding sleeping in the left lateral decubitus position is suggested [56, 57].

Pain

Pain of both cardiac and non-cardiac origin is prevalent among patients with heart failure [13, 58, 59]. The prevalence of pain also increases with worsening functional class [60]. Although pain is not a typical symptom of heart failure per se, approximately 75 % of patients experience various types of pain. Consequently, regular assessment of pain at end-of-life is warranted including documentation of nonverbal signs (grimacing, decreased functional status) among residents with cognitive dysfunction. Importantly, pain may present atypically in older adults manifested by changes such as newly developing confusion, fatigue, or depression [61].

Treatment varies with the cause of the pain. Treating underlying comorbid illness such as degenerative joint disease, arthritis, and angina is generally effective for most patients [13]. However, non-steroidal anti-inflammatory drugs should be avoided due to risk of fluid retention, negative effects on renal function and the potential for gastrointestinal bleeding [5]. For moderate to severe pain, short acting opiates initially can be used as described for dyspnea. Long-acting opiates can be used after the short-acting opiates are up titrated to a satisfactory level to manage pain [3].

Fatigue

Fatigue is one of the most common symptoms and one of the most distressing to patients with heart failure [13, 14]. The subjective sensation of fatigue ranges from lack of energy to persistent tiredness to profound exhaustion [62]. Fatigue is non-specific in that this symptom is associated with chronic illness (e.g. cancer, chronic obstructive lung disease), benign causes and more serious acute illness (e.g. acute coronary syndrome) [63]. Further, it can result from physical as well as mental illness (e.g. depression). The symptom of fatigue has implications beyond quality of life and functional capacity. Among patients with heart failure and reduced ejection fraction, both severe exertional fatigue (HR=2.59, 95 % CI: 1.09–6.16, $p=0.03$) and severe general fatigue (HR=3.20, 95 % CI: 1.62–6.31, $p=0.001$) predicted increased risk of mortality [64]. Interventions for fatigue begin with assessment and identification of any underlying cause whether physiological, psychological, or situational. Treatable causes of fatigue include but are not limited to anemia, infection, dehydration, electrolyte imbalance, thyroid dysfunction, depression, polypharmacy and pain [65]. Obtaining palliative care consultation is reasonable if a likely source of fatigue remains elusive [66].

Depression

Depression is common among patients with heart failure with approximately 20 % of patients having a major depressive disorder [67]. Quality of life is negatively affected as depression also is associated with increased hospitalization rates and higher symptom burden [43, 68, 69]. Identification of depression in the SNF population is challenging. Diagnosis is complicated by comorbid illness in some instances and cognitive deficits in others. Screening using mood is recommended over physical symptoms in this population as physical symptoms such as loss of appetite or insomnia are common in advanced heart failure [25, 70].

Considering the relationship between symptoms of heart failure and depression, careful guideline-based management of heart failure is the first step in mitigating depressive symptoms. Other interventions commonly used for depression or anxiety (e.g. psychotherapy, antidepressant medication, exercise) may be considered. Sertraline, a selective serotonin reuptake inhibitor tested in a randomized placebo-control trial of a population generally younger than found in long-term care, was safe although investigators initially reported no differences in efficacy between sertraline and placebo [71, 72]. In a subsequent analysis using growth mixture modeling, sertraline was more effective than placebo in 20 % of the responders [71]. Tricyclic antidepressants are avoided in patients with heart failure due to quinidine-like effects [3, 73]. Interventions such as exercise and psychotherapy should be considered when feasible at end-of-life.

Grief and Bereavement Management

Palliative care interventions are aimed at improving quality of life for residents by relieving distressing symptoms experienced over the disease trajectory to end-of-life. In addition to medical and nursing management of symptoms, the components of palliative care for residents and their caregivers include: (1) coordination of care using interdisciplinary care teams, (2) psychosocial and spiritual support along the disease continuum to end-of-life, (3) use of a palliative care specialist alongside medical care (integrated care delivery), and (4) inclusion of families and caregivers in all plans of care including transitioning to hospice care, grief and bereavement. Transitioning to hospice care is a mechanism for improving end-of-life care, however evidence indicates that patients with heart failure residing in long term care are less likely to enter hospice (OR=0.613, CI: 0.477–0.787) [74].

Palliative care interventions help residents' families cope with the resident's illness, their personal bereavement, and sense of loss [75]. Often, the terms grief and bereavement are used interchangeably, but are conceptually different. Grief is a process and describes a person's physical, emotional, cognitive, social and spiritual response to loss. Bereavement is the time period after the loss during which grief is experienced [76]. Bereavement outcomes vary among individuals. A good bereavement outcome is exemplified by lessening of grief over time and a return to usual life activities. Poor bereavement outcomes are described as little or no decrease in the intensity of grief after a 2 year grieving period. The chronic and progressive nature of heart failure may increase risk of a poor bereavement outcome if there is prolonged anticipatory grieving. Risk is increased particularly if their loved one also has cognitive dysfunction or dementia [77]. Other risk factors for poor bereavement outcomes that may relate to families of nursing home residents are outlined in Table 8.3. Referrals to additional resources such as grief counseling are appropriate for those at risk for poor bereavement outcomes.

Bereavement interventions generally are provided by front-line caregivers (Registered Nurses, nurse's aides, social workers, chaplains, volunteers) trained in the various aspects of bereavement. Bereavement support includes clear information about the dying process, attention and active listening by the staff, and assistance obtaining end-of-life services. Long-term care residents without dementia and their families may benefit from dignity therapy during the course of palliative care [78, 79]. Dignity therapy is a brief psychotherapeutic technique which uses reminis-

Table 8.3 Risk factors for poor bereavement outcomes

Risk factors
Ambivalent relationship with deceased person
Poor family coping (lack of cohesion, high conflict)
History of mental illness of the bereaved person
Prolonged care giving
Lack of social support

cence to record and document what is important to the individual. (See also Chap. 10.) After the document is edited to the satisfaction of the resident, it can be shared with loved ones if desired. The document potentially may be a useful bereavement tool for the family, but should be used based on individual preference. Negative response to the dignity document has been reported by those with higher caregiver burden or negative relationships with the patient [78].

Models of Care Delivery

Models for delivery of nonhospice palliative and hospice care in SNF care include in-house programs with specially trained staff, partnerships with external providers, and external consultation teams [80–82]. Improved management of symptoms, emotional and spiritual support, avoidance of hospitalization and family satisfaction are the goals of care. However, both the model for delivery and ensuing quality of symptom management are variable across facilities [44, 81]. Several factors are reported to be associated with in-house specialized training for palliative and hospice care:

- Private not-for-profit nursing homes,
- Nursing homes in the southern part of the United States,
- Nursing homes with American College of Health Care certified administrators
- Nursing homes with pain management training were nearly six times more likely to also have palliative/hospice training programs in place [81].

Training for pain management produced the largest effect on palliative care and hospice specialty training compared with nursing homes without such training. Education of healthcare providers, nursing home residents and their surrogate increases the use of hospice care and subsequently, family satisfaction with end-of-life care [83]. Casarett and colleagues conducted a randomized controlled trial testing a communication intervention for nursing home residents deemed appropriate for hospice care [83]. Criteria for determining whether the resident was an appropriate candidate for hospice referral are outlined in Table 8.4. If a resident met all three criteria, a fax was sent to the physician describing the study, possible appropriateness of hospice enrollment, and request for a reply regarding the physician's assessment of prognosis. If prognosis was determined to be 6 months or less a hospice referral ensued. The intervention resulted in increased access to hospice care and less frequent transition to acute care. Barriers to quality end-of-life care in the long-term care setting include both system and healthcare provider factors:

- Lack of palliative care training
- Low staffing
- High staff turnover
- Poor communication between healthcare providers
- Regulations that focus on restorative care [84, 85]

Table 8.4 Criteria for considering hospice referral

1. Comfort is resident or surrogate's goal of care
2. Refused both cardiopulmonary resuscitation and mechanical ventilation
3. Identified at least one need for palliative care

Most direct care providers in long-term care (registered nurses, certified nursing aides) do not receive palliative care training as part of their formal education. Therefore, support and time for training outside the academic setting is generally needed. Furthermore, nursing aides spend significantly more direct care time with nursing home residents compared with licensed personnel. Training nursing aides to assess and communicate changes in symptom status supports quality end-of-life care [85]. Initiating and sustaining in-house palliative and hospice care programs requires administrative commitment and investment in training of staff on a continuing basis, particularly with respect to high staff turnover.

Conclusion

As one in four older adults die in long-term care, planning for end-of-life is an important component of quality care in SNF. An interdisciplinary approach using the skills of nursing, medicine, social work, chaplains and volunteers is recommended for resident and family satisfaction with end-of-life. Advanced care planning and symptom management are central to achieving a good death as end-of-life is associated with a high prevalence of symptoms associated with cardiovascular disease and comorbid illness.

References

1. Mitchell SL, Teno JM, Miller SC, Mor V. A national study of the location of death for older persons with dementia. J Am Geriatr Soc. 2005;53:299–305.
2. Weitzen S, Teno JM, Fennell M, Mor V. Factors associated with site of death: a national study of where people die. Med Care. 2003;41:323–35.
3. Adler ED, Goldfinger JZ, Kalman J, Park ME, Meier DE. Palliative care in the treatment of advanced heart failure. Circulation. 2009;120:2597–606.
4. Lorenz KA, Lynn J, Dy SM, Shugarman LR, Wilkinson A, Mularski RA, et al. Evidence for improving palliative care at the end of life: a systematic review. Ann Intern Med. 2008;148:147–59.
5. Qaseem A, Snow V, Shekelle P, Casey Jr DE, Cross Jr JT, Owens DK, et al. Evidence-based interventions to improve the palliative care of pain, dyspnea, and depression at the end of life: a clinical practice guideline from the American College of Physicians. Ann Intern Med. 2008;148:141–6.
6. Kasper J, O'Malley M. Changes in characteristics, needs, and payments for elderly nursing home residents: 1999 to 2004. Washington, DC: Kaiser Family Foundation; 2007.

7. Ahmed AA, Hays CI, Liu B, Aban IB, Sims RV, Aronow WS, et al. Predictors of in-hospital mortality among hospitalized nursing home residents: an analysis of the National Hospital Discharge Surveys 2005–2006. J Am Med Dir Assoc. 2010;11:52–8.
8. Jones AL, Dwyer LL, Bercovitz AR, Strahan GW. The National Nursing Home Survey: 2004 overview. Vital Health Stat. 2009;13:1–155.
9. Yancy CW, Jessup M, Bozkurt B, Butler J, Casey Jr DE, Drazner MH, et al. 2013 ACCF/AHA guideline for the management of heart failure: a report of the American College of Cardiology Foundation/American Heart Association Task Force on Practice Guidelines. Circulation. 2013;128:e240–327.
10. Bui AL, Horwich TB, Fonarow GC. Epidemiology and risk profile of heart failure. Nat Rev Cardiol. 2011;8:30–41.
11. Funk M, Winkler CG. Epidemiology of heart failure. In: Moser DK, Riegel B, editors. Cardiac nursing: a companion to Braunwald's heart disease. St. Louis: Saunders Elsevier; 2008. p. 31–44.
12. Ekman I, Cleland JG, Andersson B, Swedberg K. Exploring symptoms in chronic heart failure. Eur J Heart Fail. 2005;7:699–703.
13. Nordgren L, Sorensen S. Symptoms experienced in the last six months of life in patients with end-stage heart failure. Eur J Cardiovasc Nurs. 2003;2:213–7.
14. Zambroski CH, Moser DK, Bhat G, Ziegler C. Impact of symptom prevalence and symptom burden on quality of life in patients with heart failure. Eur J Cardiovasc Nurs. 2005; 4:198–206.
15. Bekelman DB, Havranek EP, Becker DM, Kutner JS, Peterson PN, Wittstein IS, et al. Symptoms, depression, and quality of life in patients with heart failure. J Card Fail. 2007;13:643–8.
16. Oliver DP, Porock D, Zweig S. End-of-life care in U.S. nursing homes: a review of the evidence. J Am Med Dir Assoc. 2005;6:S21–30.
17. Goodlin SJ, Hauptman PJ, Arnold R, Grady K, Hershberger RE, Kutner J, et al. Consensus statement: palliative and supportive care in advanced heart failure. J Card Fail. 2004;10:200–9.
18. Haga K, Murray S, Reid J, Ness A, O'Donnell M, Yellowlees D, et al. Identifying community based chronic heart failure patients in the last year of life: a comparison of the Gold Standards Framework Prognostic Indicator Guide and the Seattle Heart Failure Model. Heart. 2012;98:579–83.
19. Nutter AL, Tanawuttiwat T, Silver MA. Evaluation of 6 prognostic models used to calculate mortality rates in elderly heart failure patients with a fatal heart failure admission. Congest Heart Fail. 2010;16:196–201.
20. Levy WC, Mozaffarian D, Linker DT, Sutradhar SC, Anker SD, Cropp AB, et al. The Seattle Heart Failure Model: prediction of survival in heart failure. Circulation. 2006;113:1424–33.
21. Thomas K. The GSF prognostic indicator guidance. 4th ed. 2011. http://www.goldstandardsframework.org.uk/cd-content/uploads/files/General%20Files/Prognostic%20Indicator%20Guidance%20October%202011.pdf.
22. Hogg KJ, Jenkins SM. Prognostication or identification of palliative needs in advanced heart failure: where should the focus lie? Heart. 2012;98:523–4.
23. Murray SA, Kendall M, Boyd K, Sheikh A. Illness trajectories and palliative care. BMJ. 2005;330:1007–11.
24. Gott M, Small N, Barnes S, Payne S, Seamark D. Older people's views of a good death in heart failure: implications for palliative care provision. Soc Sci Med. 2008;67:1113–21.
25. Stuart B. Palliative care and hospice in advanced heart failure. J Palliat Med. 2007;10:210–28.
26. Teno JM, Clarridge BR, Casey V, Welch LC, Wetle T, Shield R, et al. Family perspectives on end-of-life care at the last place of care. JAMA. 2004;291:88–93.
27. Wensing M, Elwyn G, Edwards A, Vingerhoets E, Grol R. Deconstructing patient centred communication and uncovering shared decision making: an observational study. BMC Med Inform Decis Mak. 2002;2:2.
28. Charles C, Gafni A, Whelan T. Shared decision-making in the medical encounter: what does it mean? (or it takes at least two to tango). Soc Sci Med. 1997;44:681–92.

29. Frank RK. Shared decision making and its role in end of life care. Br J Nurs. 2009;18:612–8.
30. Smith L, O'Sullivan P, Lo B, Chen H. An educational intervention to improve resident comfort with communication at the end of life. J Palliat Med. 2013;16:54–9.
31. Weiner JS, Cole SA. Three principles to improve clinician communication for advance care planning: overcoming emotional, cognitive, and skill barriers. J Palliat Med. 2004;7:817–29.
32. Boyd KJ, Murray SA, Kendall M, Worth A, Frederick Benton T, Clausen H. Living with advanced heart failure: a prospective, community based study of patients and their carers. Eur J Heart Fail. 2004;6:585–91.
33. Jaarsma T, Beattie JM, Ryder M, Rutten FH, McDonagh T, Mohacsi P, et al. Palliative care in heart failure: a position statement from the palliative care workshop of the Heart Failure Association of the European Society of Cardiology. Eur J Heart Fail. 2009;11:433–43.
34. Jones AL, Moss AJ, Harris-Kojetin LD. Use of advance directives in long-term care populations. NCHS Data Brief. 2011;1–8.
35. Lampert R. Quality of life and end-of-life issues for older patients with implanted cardiac rhythm devices. Clin Geriatr Med. 2012;28:693–702.
36. Fischberg D, Bull J, Casarett D, Hanson LC, Klein SM, Rotella J, et al. Five things physicians and patients should question in hospice and palliative medicine. J Pain Symptom Manage. 2013;45:595–605.
37. Goldstein NE, Lampert R, Bradley E, Lynn J, Krumholz HM. Management of implantable cardioverter defibrillators in end-of-life care. Ann Intern Med. 2004;141:835–8.
38. Levy CR, Fish R, Kramer A. Do-not-resuscitate and do-not-hospitalize directives of persons admitted to skilled nursing facilities under the Medicare benefit. J Am Geriatr Soc. 2005;53:2060–8.
39. Pastor DK, Moore G. Uncertainties of the heart: palliative care and adult heart failure. Home Healthc Nurse. 2013;31:29–36, quiz 37–8.
40. Teno JM, Shu JE, Casarett D, Spence C, Rhodes R, Connor S. Timing of referral to hospice and quality of care: length of stay and bereaved family members' perceptions of the timing of hospice referral. J Pain Symptom Manage. 2007;34:120–5.
41. Green E, Gardiner C, Gott M, Ingleton C. Exploring the extent of communication surrounding transitions to palliative care in heart failure: the perspectives of health care professionals. J Palliat Care. 2010;27:107–16.
42. Rogers AE, Addington-Hall JM, Abery AJ, McCoy AS, Bulpitt C, Coats AJ, et al. Knowledge and communication difficulties for patients with chronic heart failure: qualitative study. BMJ. 2000;321:605–7.
43. Lemond L, Allen LA. Palliative care and hospice in advanced heart failure. Prog Cardiovasc Dis. 2011;54:168–78.
44. Hanson LC, Eckert JK, Dobbs D, Williams CS, Caprio AJ, Sloane PD, et al. Symptom experience of dying long-term care residents. J Am Geriatr Soc. 2008;56:91–8.
45. Lindenfeld J, Albert NM, Boehmer JP, Collins SP, Ezekowitz JA, Givertz MM, et al. HFSA 2010 comprehensive heart failure practice guideline. J Card Fail. 2010;16:e1–194.
46. Lenz ER, Pugh LC, Milligan RA, Gift A, Suppe F. The middle-range theory of unpleasant symptoms: an update. ANS Adv Nurs Sci. 1997;19:14–27.
47. Lenz ER, Suppe F, Gift AG, Pugh LC, Milligan RA. Collaborative development of middle-range nursing theories: toward a theory of unpleasant symptoms. ANS Adv Nurs Sci. 1995;17:1–13.
48. Jurgens CY. Dyspnea. In: Moser DK, Riegel B, editors. Cardiac nursing: a companion to Brauwald's heart disease. Philadelphia: Saunders; 2008. p. 773–81.
49. Abernethy AP, Currow DC, Frith P, Fazekas BS, McHugh A, Bui C. Randomised, double blind, placebo controlled crossover trial of sustained release morphine for the management of refractory dyspnoea. BMJ. 2003;327:523–8.
50. Currow DC, Quinn S, Greene A, Bull J, Johnson MJ, Abernethy AP. The longitudinal pattern of response when morphine is used to treat chronic refractory dyspnea. J Palliat Med. 2013;16:881–6.
51. Clemens KE, Quednau I, Klaschik E. Use of oxygen and opioids in the palliation of dyspnoea in hypoxic and non-hypoxic palliative care patients: a prospective study. Support Care Cancer. 2009;17:367–77.

52. Oxberry SG, Lawrie I. Symptom control and palliative care: management of breathlessness. Br J Hosp Med (Lond). 2009;70:212–6.
53. Galbraith S, Fagan P, Perkins P, Lynch A, Booth S. Does the use of a handheld fan improve chronic dyspnea? A randomized, controlled, crossover trial. J Pain Symptom Manage. 2010;39:831–8.
54. Parshall MB, Schwartzstein RM, Adams L, Banzett RB, Manning HL, Bourbeau J, et al. An official American Thoracic Society statement: update on the mechanisms, assessment, and management of dyspnea. Am J Respir Crit Care Med. 2012;185:435–52.
55. Zambroski CH, Bekelman DB. Palliative symptom management in patients with heart failure. Prog Palliat Care. 2008;16:241–9.
56. Leung RS, Bowman ME, Parker JD, Newton GE, Bradley TD. Avoidance of the left lateral decubitus position during sleep in patients with heart failure: relationship to cardiac size and function. J Am Coll Cardiol. 2003;41:227–30.
57. Palermo P, Cattadori G, Bussotti M, Apostolo A, Contini M, Agostoni P. Lateral decubitus position generates discomfort and worsens lung function in chronic heart failure. Chest. 2005;128:1511–6.
58. Goodlin SJ, Wingate S, Albert NM, Pressler SJ, Houser J, Kwon J, et al. Investigating pain in heart failure patients: the pain assessment, incidence, and nature in heart failure (PAIN-HF) study. J Card Fail. 2012;18:776–83.
59. Shah AB, Udeoji DU, Baraghoush A, Bharadwaj P, Yennurajalingam S, Schwarz ER. An evaluation of the prevalence and severity of pain and other symptoms in acute decompensated heart failure. J Palliat Med. 2013;16:87–90.
60. Evangelista LS, Sackett E, Dracup K. Pain and heart failure: unrecognized and untreated. Eur J Cardiovasc Nurs. 2009;8:169–73.
61. Brown JA, Von Roenn JH. Symptom management in the older adult. Clin Geriatr Med. 2004;20:621–40, v–vi.
62. Dittner AJ, Wessely SC, Brown RG. The assessment of fatigue: a practical guide for clinicians and researchers. J Psychosom Res. 2004;56:157–70.
63. Friedman MM, Stephens SA. Fatigue. In: Moser DK, Riegel B, editors. Cardiac nursing: a companion to Braunwald's heart disease. Philadelphia: Saunders; 2008. p. 782–8.
64. Smith OR, Kupper N, de Jonge P, Denollet J. Distinct trajectories of fatigue in chronic heart failure and their association with prognosis. Eur J Heart Fail. 2010;12:841–8.
65. Fishbain DA, Lewis J, Cole B, Cutler B, Smets E, Rosomoff H, et al. Multidisciplinary pain facility treatment outcome for pain-associated fatigue. Pain Med. 2005;6:299–304.
66. Levy C, Morris M, Kramer A. Improving end-of-life outcomes in nursing homes by targeting residents at high-risk of mortality for palliative care: program description and evaluation. J Palliat Med. 2008;11:217–25.
67. Rutledge T, Reis VA, Linke SE, Greenberg BH, Mills PJ. Depression in heart failure a meta-analytic review of prevalence, intervention effects, and associations with clinical outcomes. J Am Coll Cardiol. 2006;48:1527–37.
68. Sullivan M, Simon G, Spertus J, Russo J. Depression-related costs in heart failure care. Arch Intern Med. 2002;162:1860–6.
69. Jiang W, Alexander J, Christopher E, Kuchibhatla M, Gaulden LH, Cuffe MS, et al. Relationship of depression to increased risk of mortality and rehospitalization in patients with congestive heart failure. Arch Intern Med. 2001;161:1849–56.
70. Williams Jr JW, Noel PH, Cordes JA, Ramirez G, Pignone M. Is this patient clinically depressed? JAMA. 2002;287:1160–70.
71. Kuchibhatla MN, Fillenbaum GG. Trajectory classes of depression in a randomized depression trial of heart failure patients: a reanalysis of the SADHART-CHF trial. Am J Geriatr Pharmacother. 2011;9:483–94.
72. O'Connor CM, Jiang W, Kuchibhatla M, Silva SG, Cuffe MS, Callwood DD, et al. Safety and efficacy of sertraline for depression in patients with heart failure: results of the SADHART-CHF (Sertraline Against Depression and Heart Disease in Chronic Heart Failure) trial. J Am Coll Cardiol. 2010;56:692–9.

73. American Geriatrics Society 2012 Beers Criteria Update Expert Panel. American Geriatrics Society updated Beers Criteria for potentially inappropriate medication use in older adults. J Am Geriatr Soc. 2012;60:616–31.
74. Zheng NT, Mukamel DB, Caprio TV, Temkin-Greener H. Hospice utilization in nursing homes: association with facility end-of-life care practices. Gerontologist. 2013;53:817–27.
75. Hall S, Kolliakou A, Petkova H, Froggatt K, Higginson IJ. Interventions for improving palliative care for older people living in nursing care homes. Cochrane Database Syst Rev. 2011;(3):CD007132.
76. Buglass E. Grief and bereavement theories. Nurs Stand. 2010;24:44–7.
77. Davidson KM. Evidence-based protocol. Family bereavement support before and after the death of a nursing home resident. J Gerontol Nurs. 2003;29:10–8.
78. Goddard C, Speck P, Martin P, Hall S. Dignity therapy for older people in care homes: a qualitative study of the views of residents and recipients of 'generativity' documents. J Adv Nurs. 2013;69:122–32.
79. Hall S, Goddard C, Opio D, Speck P, Higginson IJ. Feasibility, acceptability and potential effectiveness of Dignity Therapy for older people in care homes: a phase II randomized controlled trial of a brief palliative care psychotherapy. Palliat Med. 2012;26:703–12.
80. Miller SC. A model for successful nursing home-hospice partnerships. J Palliat Med. 2010;13:525–33.
81. Miller SC, Han B. End-of-life care in U.S. nursing homes: nursing homes with special programs and trained staff for hospice or palliative/end-of-life care. J Palliat Med. 2008;11:866–77.
82. National Quality Forum. A national framework and preferred practice for palliative and hospice care quality: a consensus report. Washington, DC: National Quality Forum; 2006.
83. Casarett D, Karlawish J, Morales K, Crowley R, Mirsch T, Asch DA. Improving the use of hospice services in nursing homes: a randomized controlled trial. JAMA. 2005;294:211–7.
84. Miller SC, Teno JM, Mor V. Hospice and palliative care in nursing homes. Clin Geriatr Med. 2004;20:717–34, vii.
85. Zheng NT, Temkin-Greener H. End-of-life care in nursing homes: the importance of CNA staff communication. J Am Med Dir Assoc. 2010;11:494–9.

Chapter 9
End-of-Life Care in Pediatric and Congenital Heart Disease

Adrienne H. Kovacs, Anne I. Dipchand, Matthias Greutmann, and Daniel Tobler

Abstract Marked advances in the diagnosis and treatment of congenital heart disease (CHD) and other pediatric-onset cardiac conditions have significantly increased survival, such that the majority of patients are now expected to reach adulthood. However, childhood mortality will not completely disappear and there continues to be a subgroup of patients with complex heart disease who will die in childhood. In addition, CHD, particularly of moderate to great complexity, remains a chronic medical condition associated with premature mortality in adulthood. Patients with congenital and pediatric heart disease have end-of-life (EOL) care considerations unique from those of adult patients dying of acquired heart disease. Though EOL considerations certainly vary between neonates, children and adolescents, and adults, an emphasis on communication and interdisciplinary care is recommended in order to improve advance care planning and EOL care for this deserving group of patients.

Keywords Congenital heart disease • Pediatric heart disease • End of life • Death and dying • Advance care planning • Palliative care

---

A.H. Kovacs, PhD, CPsych (✉)
Toronto Congenital Cardiac Centre for Adults, Peter Munk Cardiac Centre, University Health Network, 585 University Ave, 5-NU-523, Toronto, ON M5G 2N2, Canada
e-mail: adrienne.kovacs@uhn.ca

A.I. Dipchand, MD
Heart Transplant Program, Division of Cardiology, Hospital for Sick Children, Toronto, ON, Canada

M. Greutmann, MD
Adult Congenital Cardiology Program, Cardiology Department, University Hospital Zurich, Zurich, Switzerland

D. Tobler, MD
Cardiology Department, University Hospital Basel, Basel, Switzerland

S.J. Goodlin, M.W. Rich (eds.), *End-of-Life Care in Cardiovascular Disease*, DOI 10.1007/978-1-4471-6521-7_9

Key Points

- Due to significant diagnostic and treatment advances, the majority of patients with pediatric and congenital heart disease now survive to reach adulthood though morbidity and premature mortality in adulthood are common
- Patients with congenital and pediatric-onset heart disease have unique end-of-life considerations given their younger age as well as the chronic nature of their disease
- Until recently, there has been minimal research directed toward the advance care planning and end-of-life care needs of patients with pediatric and congenital heart disease
- Clinicians and researchers are encouraged to target their efforts to improve advance care planning and end-of-life care for this deserving group of patients

Introduction

Marked advances in the diagnosis and treatment of congenital heart disease (CHD) and other pediatric-onset cardiac conditions have significantly increased survival, such that the majority of patients are now expected to reach adulthood [1–3]. As an illustration, there are now more adults than children living with CHD [4]. Notwithstanding this laudable improvement in survival, there remains a subgroup of patients who will die from heart disease in childhood. In addition, adults with CHD, particularly patients with lesions of moderate to great complexity, remain at risk of premature death in adulthood. There are currently no scientific statements or consensus documents for the end-of-life (EOL) care of pediatric or congenital heart disease patients. Guidelines for adults and transitioning adolescents with CHD, however, encourage providers to incorporate a discussion of EOL issues and advance directives into routine care [5, 6].

This chapter will begin with a review of the shifting mortality patterns of patients with CHD and other pediatric-onset cardiac conditions. We will then review what is known about the EOL experiences and care strategies for pediatric cardiac patients and their families. Next, we will summarize EOL experiences and care strategies for adults with CHD. The chapter will conclude with clinical recommendations to improve the EOL care of patients with congenital or pediatric-onset heart disease across the lifespan.

Congenital and Pediatric-Onset Heart Disease: Prevalence and Mortality

Congenital malformations and diseases of the heart are the fifth and sixth leading causes of death for children in the United States (following only accidents, homicide, suicide, and malignant neoplasms) [7]. Cardiac defects are the most common congenital defect, affecting almost 1 in every 100 newborns [8]. In the mid twentieth century, only 25 % of infants born with CHD survived to reach adulthood, but it is now estimated that approximately 90 % will reach this important milestone [9]. This striking improvement reflects significant advances in diagnostic capabilities (including fetal diagnosis), surgical and percutaneous intervention techniques, and critical care management. Oster et al. reported temporal trends in survival among infants with critical CHD included within the Metropolitan Atlanta Congenital Defects Program [10]. They found that 1-year survival increased from 67 % for infants born in 1979–1993 to 83 % for infants born between 1994 and 2005. As a group, patients born with CHD are clearly living longer than ever before [11]. An analysis of the Center for Disease Control Multiple Cause-of-Death registry from 1979 to 2005 revealed significant reductions in child and adult death rates among many forms of CHD including transposition of the great arteries (71 % reduction), coarctation of the aorta (70 % reduction), and tetralogy of Fallot (50 % reduction) [12]. A review of United States death certificates demonstrated a 24 % decline in mortality due to CHD from 1999 to 2006; further, for patients surviving the first year of life, 76 % of deaths occurred in adulthood [3]. Similarly, a population-based cohort study in Quebec, Canada revealed a significant reduction in mortality rates among all age groups below 65 years with CHD, although the largest mortality reduction was among the infant subgroup [2].

Due to this shift in mortality, the majority of deaths due to CHD now occur in adults [2, 3]. Two important facts remain, however:

1. Despite the success in treating children with complex CHD, childhood mortality will not completely disappear and there continues to be a subgroup of patients with complex CHD who will die in childhood.
2. CHD, particularly of moderate to great complexity, remains a chronic medical condition impacting health-related quality of life during adulthood.

It is important to recognize that the majority of patients among these rapidly growing novel adult patient cohorts are not cured. Many of these young adults with complex CHD remain at risk of premature mortality. The average lifespan across all forms of CHD is around 50 years of age [13], though this is certainly an evolving cohort for which longer life expectancies are indeed expected. Premature mortality is predominantly caused by cardiac-related deaths, mainly sudden death, heart failure and perioperative deaths [12–16]. Given the demographics and evolution of the adult patient cohorts, we expect a significant increase in the overall number of young adults at risk of premature death during the upcoming decades. In summary, significant improvements in survival rates do not translate into a normal life expectancy for the majority of individuals born with CHD of moderate to great complexity.

Though representing a significant proportion of pediatric cardiology patients, CHD is not the only cardiac condition affecting young people. Cardiomyopathy is far less common, but is one of the common causes of heart failure in the pediatric setting and is the most frequent reason for heart transplantation in children over 12 months of age [17]. Hypertrophic cardiomyopathy is the most common cause of sudden cardiac death in young people [18]. Another leading cause of sudden cardiac death in young people is long QT syndrome [19]. For both hypertrophic cardiomyopathy and long QT syndrome, implanted cardiac devices (implanted cardioverter defibrillators, cardiac resynchronization therapy) have extended life expectancies for many high-risk patients [18, 20].

Heart transplantation for children and adults with end-stage heart failure has been associated with improving outcomes. In pediatric transplant patients, survival beyond 20 years has now become common [21]. However, given uncertainty regarding candidacy for transplantation, survival while on the waiting list, and outcomes of higher risk transplantations, EOL discussions should be included early in this evaluation [22, 23]. For both pediatric and adult patients, a discussion of EOL decision-making should be provided concurrently with evaluation for high risk transplantation [23, 24]. Further, the option not to proceed with transplantation can also be considered an "active" and acceptable decision [23].

The Need for Specialized End-of-Life Care

Although the majority of patients with congenital or pediatric-onset heart disease reach adulthood, unique EOL care needs exist for two groups of individuals: (1) young patients (and their families) who face death in the pediatric cardiology setting and (2) patients with CHD who have reached adulthood but still face premature mortality. Both groups have EOL care considerations unique from those of adult patients dying of acquired heart disease. The special considerations regarding their EOL care are listed below:

- Pediatric patients are much younger (and might even be neonates) and their parents are typically responsible for EOL care decision-making.
- Adults with complex CHD also typically die decades younger than their counterparts with acquired heart disease, despite having "beaten the odds" many times.
- It is not uncommon to meet a 40-year old with CHD who has survived several risky surgical procedures and recalls that their parents were informed that their child was initially not expected to survive childhood, let alone reach middle-aged years.
- Adults with CHD have lived since birth with a chronic medical condition; this is certainly a distinction from adults with acquired heart disease who might have considered themselves healthy for the majority of their lives.
- The situation wherein the child or adult is not at imminent risk of dying, but a decision has been made to NOT pursue a high risk surgery or transplantation (either based on candidacy or family/patient decision-making). There are circumstances in which survival can extend for several years following this acknowledgement. Though often there has been no change in symptomatology or no change in need for medical supports or interventions, this is a significant philosophical change for the patient and the family, and EOL considerations in the early phase of this journey are unique and require addressing.

End of Life Care in Pediatric Cardiology

A complete review of pediatric palliative care is beyond the scope of this paper. Readers are directed to statements and guidelines published by the Committee on Bioethics of the American Academy of Pediatrics regarding palliative care for children, forgoing life-sustaining medical treatment, and informed consent, parental permission, and pediatric patient assent [25–27]. In 2009, the National Hospital and Palliative Care Organization published standards of practice for pediatric palliative care and hospice [28]. Readers might also be interested in online training materials offered from the Initiative for Pediatric Palliative Care (www.ippeweb.org).

Pediatric advance care planning has the same goals as adult advance care planning, namely the preservation of quality of life, provision of optimal care, and avoidance of needless suffering [29]. Planning discussions can include parents, pediatric patients (as appropriate), providers and clinical ethicists [29]. Documentation should include treatments (e.g., cardiopulmonary resuscitation, tube feeding, intubation) that are identified to be helpful or harmful, as well as the ethical and personal reasons that led to these decisions [29]. Within the field of pediatric cardiology, there are different considerations for neonates, children and adolescents, as well as for parents who are the ones primarily responsible for medical decision-making.

Neonates

Rychik described the role of palliative care in the context of a prenatal diagnosis of hypoplastic left heart syndrome (HLHS), which is one of the most complex heart defects [30]. Despite the availability of surgical interventions, some families will decide not to surgically intervene and instead choose a palliative care approach following birth [30]. Rychik emphasized the importance of objective prenatal counseling and support for the parents to arrive at the decision that is right for them. At his institution, 82 % of parents continue with pregnancy, 11 % choose termination, and 7 % of parents choose palliative care [31]. Rychik described the latter decision as "one in which parents are hoping to prevent later pain and suffering for the future child or adult" [30]. It should be noted that the rate of elective termination by parents as a management option for HLHS varies widely across centres and countries. Murtuza et al. surveyed surgeons and reported (estimated) termination rates ranging from 3 to 95 % [32]. A discussion and debate regarding the appropriateness of the palliative care option given improved HLHS survival rates has ensued [33]. When the palliative care option is chosen, death of the newborn usually occurs within 24–48 h after birth, although some infants survive for weeks [30]. This is a unique situation in which advance care planning and discussion of EOL issues may occur several weeks or even months prior to birth.

Trines et al. reported that 49 % of women with fetuses with prenatal diagnoses of severe CHD at their centre chose termination [34]. Comprehensive parental counseling includes not only information about the fetus's cardiac anatomy and perinatal and neonatal management, but also matters related to longer term mortality and morbidity (e.g., survival to adulthood, life expectancy, anticipated cardiac complications, potential neurological issues, and quality of life considerations). Parents may take any or all of those factors into account when considering management options, including termination. The absence of a "cure" for most forms of CHD differs from that of other pediatric medical conditions for which there are interventions that will result in a high likelihood of a normal adulthood and family and quality of life.

The highest mortality rates following cardiac surgery are observed in neonates [35]. Although surgical interventions for HLHS have advanced in recent years, a retrospective review of outcomes from 1983 to 2004 revealed that neonate deaths have historically been common even following surgical treatment [36]. Specifically, 82/134 (62 %) of children in this study died and the majority of these deaths occurred in the intensive care unit. This study highlighted the importance of developing strategies to support parents (and staff) at the time of HLHS diagnosis as well as before and following an infant's death. A staged approach to EOL decision-making for parents was recommended.

Palliative care can be especially effective in the neonate setting [37]. Formal palliative care consultations are associated with fewer medical procedures (e.g., blood draws, mechanical ventilation) and increased supportive services (e.g., chaplains, social services) [38].

Children and Adolescents

Oncology has led the way in palliative care and guidelines from pediatric oncology are worthwhile to consider in the absence of pediatric cardiology-specific scientific statements or consensus guidelines. We agree with our oncology colleagues in that, "it should always be possible for a terminally ill child to die without unnecessary physical pain, fear, or anxiety" [39]. Themes expressed from dying children with cancer include viewing death as a journey to a faraway place rather than an end, a preference to avoid living in pain, concerns about parental adjustment after their death, and both sadness and acceptance of impending death [40]. Despite advances in pediatric oncology palliative care, challenges do remain. For example, interviews with parents of 48 children who had died of cancer revealed that they often perceived that anxiety had not been effectively managed [41].

Honest communication between children, their parents, and providers regarding a poor prognosis is recommended at a level that is developmentally-appropriate for an individual child. A child's understanding of the concept of death develops with age [42]. It is within the 5–10 year age range that children being to understand the permanence of death; a more abstract understanding typically develops in adolescence. There are developmental, psychosocial and clinical considerations when considering the degree to which children and adolescents should be included in EOL medical decision-making [43]. Appropriate inclusion of children and adolescents respects children's capacities and the ethical principle of self-determination, and also enhances communication and adherence [43]. Though parents might legally be the ones to provide informed consent for treatment (or the decision to forego life-sustaining medical treatment), children's opinions and assent should be sought [26, 27, 43].

A study of Swedish parents whose children (aged 24 and younger) had died from cancer revealed that one-third had talked about death with their child and two-thirds had not [44]. Rates of talking about death varied according to the child's age at the time of death: 16 % in 0–4-year olds, 36 % in 5–8 year olds, 50 % in 9–15 year olds, and 41 % in 16–24 year olds. No parents who had discussed death regretted their decision, whereas one-quarter of parents who did not talk about death did regret not having this discussion. Thus, most parents, regardless of their approach, were satisfied with their decision. Regrets about not having discussed death were more common for parents who thought their children were aware of their imminent death. Study authors share an important clinical implication of their study: parents who wonder whether they should or should not talk about death with their dying children might benefit from knowing the results of this study in which no parents who had discussed death regretting having done so.

Freyer described the situation facing adolescents dying from serious health conditions as "a paradox of emerging capabilities and diminishing possibilities" [45]. A retrospective study of 103 adolescents who died from cancer revealed that EOL discussions typically occurred late in the disease, thus failing to allow adequate time for psychological preparation [46].

Adolescents should be included in discussions regarding foregoing life-prolonging treatments [47]. McAliley et al. interviewed 107 teens between the ages of 15 and 18 years of age and found that most did not feel uncomfortable discussing advance directives and thought it was somewhat or very important for them to have a living will [48]. Adolescents with HIV responded favorably to an advance care planning process that included discussion about EOL wishes with family members [49]. Advance care planning documents have been shown to be appropriate for and accepted by adolescents with life-limiting illnesses [50]. A survey of healthy adolescents and those with chronic illnesses revealed that most seek a shared approach to decision-making, but that the timing of EOL discussions varied according to medical status, with chronically ill adolescents preferring to have EOL discussions later in the disease course [51].

Parents

One unique aspect of pediatric EOL care is the degree to which parents are primarily responsible for decision-making. Gillam and Sullivan highlight the ethical considerations regarding parental involvement in EOL decision-making for infants and young children [52]. They suggest that parents should be and generally wish to be involved in decision-making and do not suffer negative psychological consequences for doing so. They also recommend, however, that the degree of parental involvement should be negotiated on an individual basis, partly because parents vary in their wishes to take final responsibility for decision-making. Though it may be difficult to inform a parent that their child is dying, this discussion best occurs earlier rather than later to avoid needless physical and emotional suffering [47].

Though the death of a child is an extremely difficult event, there is evidence to suggest that adaptation does occur. Brosig et al. conducted 19 interviews with parents whose infants (less than 12 months old) had died (due to causes including but not limited to cardiac conditions) [53]. They found that parents generally adapted positively and identified seven important aspects of care:

1. honest information from medical staff,
2. involvement in decision-making, including the withdrawal of life support,
3. care of the parents in addition to the infant,
4. comfortable hospital environment,
5. trust in nursing care,
6. presence of physicians throughout the process of the infant's death, and
7. support from other care providers, including chaplains, social workers, child life specialists, and palliative care specialists.

Effective coping strategies included family support, keeping memories of the deceased infant, spirituality/faith, altruism and "giving back" to the hospital, refocusing on life, validating their decision to withdraw life support, and bereavement support groups.

End of Life Care in Adult Congenital Cardiology

Advance Care Planning

Guidelines for the management of adults with CHD recommend that patients complete advance directives (written documents that allow individuals to make their health care wishes known before becoming very ill) "at a time during which they are not morbidly ill or hospitalized, so that they can express their wishes in a less stressful setting" [5]. This was a Class I, Level of Evidence C recommendation, such that the benefits outweigh the risks and the intervention should be performed, although the recommendation is based upon consensus opinion rather than rigorous empirical study. In a recent study of 200 adult CHD outpatients (mean age = 35 years; 48 % female), only 5 % of patients had completed advance directives [54]. Over half of the patients had no knowledge of advance directives, although the majority (87 %) thought that it would be important to have an advance directive if they were dying and unable to communicate for themselves. Less than one in five patients had formally identified a substitute decision-maker, and this did not vary by defect complexity. Thus, advance care documents are likely not being prepared as recommended and recognized as important by both providers and patients.

The study cohort of 200 adults with CHD, in addition to 48 adult CHD health providers, were surveyed regarding the frequency of EOL discussions [55]. Although only two patients recalled having EOL discussions with their providers, over half recalled discussing EOL preferences with other people, most commonly partners and parents. Interestingly, the perception for the need for and timing of EOL issues differed quite markedly between patients and providers. Independent of any medical or demographic variables, over three-quarters of patients thought that a member of their medical team should raise EOL issues. Providers, on the other hand, reported that they were more likely to have EOL discussions with patients with disease of greater complexity and shorter life expectancy. Almost twice as many patients as providers (62 % vs. 38 %) thought that EOL issues should occur earlier in the disease course and prior to the onset of life-threatening complications.

One challenge when initiating conversations about advance care planning with this patient population is that many patients anticipate living much longer than is expected of them. A study of patients with CHD between the ages of 16 and 20 years revealed that they expected to live to an average of 75 years, only 4 years less than their estimate for their healthy peers, and for the majority, much longer than was predicted by the researchers [56]. Despite this over-estimation, there is recent evidence to suggest that adults with CHD are indeed interested in becoming knowledgeable about longer term health expectations. In the aforementioned cohort of 200 adult CHD outpatients, although a minority (35 %) reported that they would want to know their individual expected life expectancy, the majority (70 %) were interested in more general information pertaining to the life expectancy associated with their form of CHD [54]. Further, when contemplating EOL issues, these patients highlighted the importance of including families in their decision-making, receiving honest answers from their doctors, and understanding their treatment choices [54]. Our own clinical experience suggests that many patients with CHD of mild or moderate complexity hold inaccurate assumptions about their own life expectancy. They often express relief after learning that they have a normal or near-normal life expectancy.

A major barrier to EOL discussions with adults with CHD is the absence of tools for reliably assessing prognosis [55, 57, 58]. Surveyed adult CHD providers reported that greater prognostic certainty would enhance advance care planning discussions [58]. In addition, both providers and patients highlighted factors related to a strong patient-provider relationship (e.g., knowing patients/doctors for a long time, having a trusting relationship) as being important facilitators of EOL discussions [58]. It has been noted that a meaningful relationship between an adult with CHD and their provider "is a critical and determining foundation upon which all other facets of effective and therapeutic clinical care and advance care planning strategies are built" [59].

End-of-Life Care

Matters related to death and dying are known to be important to many adults with CHD [60]. However, death in young adults is unusual for many cardiologists and adults with CHD may thus be at risk of receiving aggressive medical care prior to death, even in the face of medical futility. A retrospective study of 48 adults with CHD (mean age = 37 years) who died while admitted to a tertiary hospital was undertaken [57]. Despite the fact that the median length of hospital admission prior to death was 16 days and the majority of patients were considered to have advanced disease, only five patients (10 %) had documented EOL discussions either during their final admission or a previous outpatient clinic visit. Documented EOL discussions also occurred with the substitute decision-maker of 21 patients (41 %); these discussions took place a mean of 2 days before death. Two-thirds of the patients died in the intensive care unit and almost half were on mechanical ventilation when they died. Approximately half of the patients died under attempted resuscitation, although this was less likely among patients with documented EOL discussions and did not occur at all for patients who had been referred to palliative care. This stressful situation for patients, substitute decision-makers and health care providers could likely have been ameliorated if advance care directives were completed earlier in the disease course.

Advance Care Planning and End-of-Life Care Strategies

Table 9.1 offers recommendations for advance care planning and EOL care applicable to working with patients with CHD or other pediatric-onset cardiac conditions. These are divided into two categories: (i) suggestions to optimize patient-provider communication and (ii) strategies to enhance interdisciplinary care.

Table 9.2 offers recommendations unique to the pediatric cardiology setting; strategies focus on patient-provider and parent-provider communication. Below, we highlight the importance of effective communication and suggest caution when considering the role of technology at the end of a patient's life.

Focus on Communication

Greutmann et al. offer a conceptualization of "comprehensive care" in CHD [58], which is also applicable to patients with other forms of pediatric-onset heart disease. In their model, comprehensive care begins at birth, or even earlier in the case of prenatal diagnoses. Discussions regarding prognosis occur with parents of affected children, and adolescents are gradually included into discussions and the decision-making process. Advance care planning and EOL discussions may occur more frequently and formally in the adult care setting when cardiac complications (e.g., heart failure, arrhythmias) and hospitalizations become more common. A transition from active treatment to palliative care becomes appropriate for many patients.

Effective patient-provider communication is critical when facing death, regardless of patient age. Hsiao et al. interviewed 20 child and parent pairs in which the child (mean age = 14 years) had cardiac disease or cancer with a poor prognosis [62].

Table 9.1 Recommendations for advance care planning and end-of-life care in pediatric and congenital heart disease

Category	Recommendation
Patient – provider communication	Offer conversations about EOL earlier in the disease course and to patients/parents irrespective of disease complexity
	Consider scheduling a unique visit to discuss EOL and advance care planning issues
	Normalize advance care planning discussions
	Acknowledge a patient-reported history of previously having “beaten the odds”
	Repeat discussions as necessary (e.g., after a significant change in clinical status)
	Provide information in the simplest and clearest language possible
	Ensure sensitivity to a patient’s cultural background
	Be aware of the cognitive and developmental abilities of an individual patient and tailor EOL discussions accordingly
	Provide information in verbal and written format (consider developing education materials specific to advance care planning for patients with congenital or pediatric heart disease)
	Offer a range of time regarding prognosis and acknowledge uncertainty
	Ensure a clear discussion of the expected outcomes of cardiopulmonary resuscitation
	For patients with ICDs, discuss the implications of device deactivation
	Facilitate the completion of advance care planning documents (e.g., naming of substitute decision-maker)
	Emphasize that the team will continue to provide care in the event of a decision to cease life-sustaining medical treatment
	Provide ongoing support and communication following a referral to palliative care
	After a patient’s death, follow up with a telephone call and/or condolence card
Interdisciplinary care	Consider identifying a primary attending physician for an inpatient facing death who can provide continuity of care (and communication) in settings in which attending staff typically rotates
	Ensure that EOL discussions are well-documented so that all members of the outpatient and inpatient care team are informed
	Include health providers on the care team who can attend to the physical, emotional, and spiritual needs of patients facing death
	Consider referrals to palliative care
	Consider bioethics consultations in challenging situations
	As necessary, use interpreters to ensure that information is provided in the patient’s language
	When an adult with CHD dies, consider informing that patient’s pediatric providers
	Acknowledge emotional reactions within members of the team when a (young) patient dies and encourage self-care practices

Note: Recommendations drawn from a number of sources, including [6, 25, 47, 55, 59, 61]

Table 9.2 Recommendations for advance care planning and end-of-life care specific to the pediatric cardiology setting

Category	Recommendation
Patient – provider communication	Do not exclude children from discussions of health and illness
	Do not assume that children are unaware of their poor health
	Provide the pediatric patient with information regarding their medical condition, treatment options, and prognosis
	Adapt communication to the chronological age and developmental level of the patient
	Give children the opportunity to ask questions (with and without their parents present)
	Encourage children and adolescent to participate in medical decision-making to the extent they are able and also desire to be involved
	Be mindful of the age of consent, which varies across states and provinces
Parent – provider communication	Unless parents suspect their children do not want to talk about death, they can be encouraged to maintain an open dialogue about death and dying
	Develop a system by which interested parents can connect with other parents who are going through (or have gone through) a similar situation
	Address parental anticipatory bereavement and grief
	Consider offering parents the option to be present during any resuscitation efforts

Note: Recommendations drawn from a number of sources, including [25, 47, 61]

The following five domains of effective physician communication were identified by both patients and parents:

1. strong relationship building skills that facilitated rapport and trust,
2. demonstration of knowledge, competence, and effort,
3. effective information exchange,
4. availability of physician to child and parent, and
5. appropriate level of child and parent involvement in discussions and decision-making.

Adults with CHD have similarly identified honest communication as a priority for effective EOL care [54].

Advance care planning and EOL discussions should always be tailored to the cognitive and developmental abilities of an individual patient. Further, it is wise to ensure comprehension of medical decisions and EOL discussions by asking patients to repeat a summary of what was said or decided. Such practices are likely even more important when working with patients with CHD in the pediatric or adult setting, because individuals born with CHD are at increased risk of cognitive and developmental disorders [63]. The term "life-shortening medical condition" can be a helpful term for many patients [59]. It is also important to attend to the potential religious and spiritual facets of death, dying, burial or cremation, and mourning for many individuals. Guidance from spiritual care or religious leaders can be extremely helpful to guide the timing and context of many EOL interventions.

The specific components of EOL discussions will vary depending upon a patient's unique cardiac condition. For example, for patients at risk of arrhythmic death (e.g., patients with inherited arrhythmias, hypertrophic cardiomyopathy, certain forms of CHD), a discussion of the anticipated outcomes of cardiopulmonary resuscitation is critical. For pediatric patients listed for heart transplantation, the decision not to proceed with resuscitation is not uncommon [23]. A discussion of the medical, ethical, and legal implications of device deactivation is also appropriate for patients with implantable cardioverter defibrillators (ICDs) [20, 64, 65].

Given obvious room for improvement in the EOL care of adults with CHD, it is a positive finding that the majority of surveyed adult CHD providers want more information and resources about advance care planning and communication strategies for EOL or advance care planning [58]. The "Ask-Tell-Ask" framework [66] has been recommended for use with adults with CHD [59]. It is also applicable to adolescents with heart disease or family members of patients of any age.

The first step is to ASK what the patient (or parent) understands about an issue (e.g., heart disease diagnosis, treatment plans, prognosis, cardiopulmonary resuscitation) and what they wish to learn.

Second, a provider should TELL information that is requested by the patient/parent or that the provider knows should be communicated to them.

Third, the provider can again ASK for a confirmation of understanding and whether there are additional questions.

Consider the Role of Technology

In-hospital deaths among children with advanced heart disease and adults with CHD are frequently accompanied by highly technical care, including cardiopulmonary resuscitation, mechanical circulatory support and ventilation [57, 67]. Further, most in-hospital deaths occur in intensive care units [57, 67]. Guidelines for palliative care in pediatric oncology emphasize the avoidance of curative treatment after the point at which cure is no longer possible [39]. A similar approach is reasonable in the cardiac setting. The Academy of Pediatrics emphasizes that technology should only be used when the benefits outweigh the harms [25, 26], and this principle certainly holds equally true for adults. Further, at the time of evaluation for high-risk interventions, such as ventricular assist devices and/or transplantation, a concurrent palliative care approach should be undertaken [24].

A study of the in-hospital deaths of 111 children with heart disease (mean age = 4.8 months; age range = 1 day to 20.5 years; median hospital stay = 22 days) revealed that "highly technical care" was common [67]. Specifically, within 24 h of death, 92 % received ventilation, 46 % had extracorporeal membrane oxygenation (ECMO) or a ventricular assist device (VAD), 23 % had peritoneal drains, and 19 % had gastrostomy tubes. Approximately two-thirds of patients died after disease-directed interventions were withdrawn, and one-quarter died during resuscitation.

Medical records documented parental presence at bedside for 83 % of deaths. A documented EOL discussion occurred with three-quarters of patients and 16 % had received palliative care consultation. These results mirror the retrospective study of 48 adults with CHD in which documented EOL discussions with patients were rare but mechanical ventilation and resuscitation efforts were common [57]. Thus, clear discussions about the use of technology should become part of routine advance care planning discussions, particularly because the general public tends to over-estimate survival following out-of-hospital cardiac arrest and inpatient resuscitation [68, 69].

Formal palliative care consultations have been associated with fewer medical procedures (including cardiopulmonary resuscitation attempts) in neonates, children, and adults [38, 57, 70]. Unfortunately, palliative care referrals remain uncommon, especially for neonates and children dying from circulatory disease [28].

Conclusions

Though the basic principles of advance care planning and EOL care apply to patients with both acquired and congenital/pediatric heart disease, there are several unique considerations for the latter group. Within the pediatric cardiology setting, a discussion of palliative care may begin as early as the prenatal period. For children and adolescents with complex heart disease, an approach to advance care planning and EOL discussions must take into account the cognitive and developmental abilities of each individual patient, while at the same time recognizing the prominent role of the parent in the decision-making process. Though the majority of infants born with CHD now reach adulthood, CHD represents a chronic life-shortening medical condition such that many patients remain at risk of morbidity and premature mortality in adulthood. Unfortunately, advance care planning and EOL discussions have not been commonly documented among adults with CHD. For clinicians in both pediatric and adult cardiology settings, an emphasis on effective communication and interdisciplinary care is recommended. Clinicians should also take care to minimize the use of unwarranted highly technical care. Further, decisions not to proceed with high-risk interventions should be supported.

There is also ample room for empirical investigation in this area. Researchers are encouraged to develop and evaluate strategies to overcome barriers and improve advance care planning and EOL discussions with children, adolescents, and adults with congenital or pediatric-onset heart disease. Such strategies may include the development of patient and family educational materials, provider communication skills training, prognostication tools, and the identification of the optimal timing of referral to and role of interdisciplinary team members including palliative care, bioethics, spiritual care, and mental health providers. The ultimate goal of combined clinical and research efforts is to improve advance care planning and EOL care for this deserving group of patients.

References

1. Greutmann M, Tobler D. Changing epidemiology and mortality in adult congenital heart disease: looking into the future. Future Cardiol. 2012;8(2):171–7. Epub 2012/03/15.
2. Khairy P, Ionescu-Ittu R, Mackie AS, Abrahamowicz M, Pilote L, Marelli AJ. Changing mortality in congenital heart disease. J Am Coll Cardiol. 2010;56(14):1149–57. Epub 2010/09/25.
3. Gilboa SM, Salemi JL, Nembhard WN, Fixler DE, Correa A. Mortality resulting from congenital heart disease among children and adults in the United States, 1999 to 2006. Circulation. 2010;122(22):2254–63. Epub 2010/11/26.
4. Marelli AJ, Mackie AS, Ionescu-Ittu R, Rahme E, Pilote L. Congenital heart disease in the general population: changing prevalence and age distribution. Circulation. 2007;115(2): 163–72. Epub 2007/01/11.
5. Warnes CA, Williams RG, Bashore TM, Child JS, Connolly HM, Dearani JA, et al. ACC/AHA 2008 Guidelines for the Management of Adults with Congenital Heart Disease: Executive Summary: a report of the American College of Cardiology/American Heart Association Task Force on Practice Guidelines (writing committee to develop guidelines for the management of adults with congenital heart disease). Circulation. 2008;118(23): 2395–451. Epub 2008/11/11.
6. Sable C, Foster E, Uzark K, Bjornsen K, Canobbio MM, Connolly HM, et al. Best practices in managing transition to adulthood for adolescents with congenital heart disease: the transition process and medical and psychosocial issues: a scientific statement from the American Heart Association. Circulation. 2011;123(13):1454–85. Epub 2011/03/02.
7. Hamilton BE, Hoyert DL, Martin JA, Strobino DM, Guyer B. Annual summary of vital statistics: 2010–2011. Pediatrics. 2013;131(3):548–58. Epub 2013/02/13.
8. van der Linde D, Konings EE, Slager MA, Witsenburg M, Helbing WA, Takkenberg JJ, et al. Birth prevalence of congenital heart disease worldwide: a systematic review and meta-analysis. J Am Coll Cardiol. 2011;58(21):2241–7. Epub 2011/11/15.
9. Warnes CA. The adult with congenital heart disease: born to be bad? J Am Coll Cardiol. 2005;46(1):1–8. Epub 2005/07/05.
10. Oster ME, Lee KA, Honein MA, Riehle-Colarusso T, Shin M, Correa A. Temporal trends in survival among infants with critical congenital heart defects. Pediatrics. 2013;131(5):e1502–8. Epub 2013/04/24.
11. Moons P, Bovijn L, Budts W, Belmans A, Gewillig M. Temporal trends in survival to adulthood among patients born with congenital heart disease from 1970 to 1992 in Belgium. Circulation. 2010;122(22):2264–72. Epub 2010/11/26.
12. Pillutla P, Shetty KD, Foster E. Mortality associated with adult congenital heart disease: trends in the US population from 1979 to 2005. Am Heart J. 2009;158(5):874–9. Epub 2009/10/27.
13. Verheugt CL, Uiterwaal CSPM, Van Der Velde ET, Meijboom FJ, Pieper PG, Van Dijk APJ, et al. Mortality in adult congenital heart disease. Eur Heart J. 2010;31(10):1220–9.
14. Nieminen HP, Jokinen EV, Sairanen HI. Causes of late deaths after pediatric cardiac surgery: a population-based study. J Am Coll Cardiol. 2007;50(13):1263–71.
15. Oechslin EN, Harrison DA, Connelly MS, Webb GD, Siu SC. Mode of death in adults with congenital heart disease. Am J Cardiol. 2000;86(10):1111–6.
16. Engelfriet P, Boersma E, Oechslin E, Tijssen J, Gatzoulis MA, Thilen U, et al. The spectrum of adult congenital heart disease in Europe: morbidity and mortality in a 5 year follow-up period. The Euro Heart Survey on adult congenital heart disease. Eur Heart J. 2005;26(21):2325–33. Epub 2005/07/06.
17. Wilkinson JD, Landy DC, Colan SD, Towbin JA, Sleeper LA, Orav EJ, et al. The pediatric cardiomyopathy registry and heart failure: key results from the first 15 years. Heart Fail Clin. 2010;6(4):401–13, vii. Epub 2010/09/28.
18. Maron BJ, Maron MS. Hypertrophic cardiomyopathy. Lancet. 2013;381(9862):242–55. Epub 2012/08/10.

19. Schwartz P. Practical issues in the management of the long QT syndrome: focus on diagnosis and therapy. Swiss Med Wkly. 2013;143:w13843. Epub 2013/10/04.
20. Kramer DB, Kesselheim AS, Salberg L, Brock DW, Maisel WH. Ethical and legal views regarding deactivation of cardiac implantable electrical devices in patients with hypertrophic cardiomyopathy. Am J Cardiol. 2011;107(7):1071–5.e5. Epub 2011/02/08.
21. Dodd DA. Pediatric heart failure and transplantation: where are we in 2013? Curr Opin Pediatr. 2013. Epub 2013/09/03.
22. Crone CC, Marcangelo MJ, Shuster Jr JL. An approach to the patient with organ failure: transplantation and end-of-life treatment decisions. Med Clin North Am. 2010;94(6):1241–54, xii. Epub 2010/10/19.
23. Dipchand AI. Decision-making in the face of end-stage organ failure: high-risk transplantation and end-of-life care. Curr Opin Organ Transplant. 2012;17(5):520–4. Epub 2012/08/15.
24. Bowater SE, Speakman JK, Thorne SA. End-of-life care in adults with congenital heart disease: now is the time to act. Curr Opin Support Palliat Care. 2013;7(1):8–13. Epub 2012/12/01.
25. American Academy of Pediatrics. Committee on Bioethics and Committee on Hospital Care. Palliative care for children. Pediatrics. 2000;106(2 Pt 1):351–7. Epub 2000/08/02.
26. American Academy of Pediatrics Committee on Bioethics. Guidelines on foregoing life-sustaining medical treatment. Pediatrics. 1994;93(3):532–6. Epub 1994/03/01.
27. Informed consent, parental permission, and assent in pediatric practice. Committee on Bioethics, American Academy of Pediatrics. Pediatrics. 1995;95(2):314–7. Epub 1995/02/01.
28. Organization NHaPC. Standards of practice for pediatric palliative care and hospice. 2009:1–36.
29. Hammes BJ, Klevan J, Kempf M, Williams MS. Pediatric advance care planning. J Palliat Med. 2005;8(4):766–73. Epub 2005/09/01.
30. Rychik J. What does palliative care mean in prenatal diagnosis of congenital heart disease? World J Pediatr Congenit Heart Surg. 2013;4(1):80–4. Epub 2013/06/27.
31. Rychik J, Szwast A, Natarajan S, Quartermain M, Donaghue DD, Combs J, et al. Perinatal and early surgical outcome for the fetus with hypoplastic left heart syndrome: a 5-year single institutional experience. Ultrasound Obstet Gynecol. 2010;36(4):465–70. Epub 2010/05/26.
32. Murtuza B, Elliott MJ. Changing attitudes to the management of hypoplastic left heart syndrome: a European perspective. Cardiol Young. 2011;21 Suppl 2:148–58. Epub 2011/12/14.
33. Ross LF, Frader J. Hypoplastic left heart syndrome: a paradigm case for examining conscientious objection in pediatric practice. J Pediatr. 2009;155(1):12–5. Epub 2009/06/30.
34. Trines J, Fruitman D, Zuo KJ, Smallhorn JF, Hornberger LK, Mackie AS. Effectiveness of prenatal screening for congenital heart disease: assessment in a jurisdiction with universal access to health care. Can J Cardiol. 2013;29(7):879–85. Epub 2013/06/04.
35. Krishnamurthy G, Ratner V, Bacha E. Neonatal cardiac care, a perspective. Semin Thorac Cardiovasc Surg Pediatr Card Surg Annu. 2013;16(1):21–31. Epub 2013/04/09.
36. Cantwell-Bartl AM, Tibballs J. Place, age, and mode of death of infants and children with hypoplastic left heart syndrome: implications for medical counselling, psychological counselling, and palliative care. J Palliat Care. 2008;24(2):76–84. Epub 2008/08/07.
37. Martin M. Missed opportunities: a case study of barriers to the delivery of palliative care on neonatal intensive care units. Int J Palliat Nurs. 2013;19(5):251–6. Epub 2013/08/27.
38. Pierucci RL, Kirby RS, Leuthner SR. End-of-life care for neonates and infants: the experience and effects of a palliative care consultation service. Pediatrics. 2001;108(3):653–60. Epub 2001/09/05.
39. Masera G, Spinetta JJ, Jankovic M, Ablin AR, D'Angio GJ, Van Dongen-Melman J, et al. Guidelines for assistance to terminally ill children with cancer: a report of the SIOP Working Committee on psychosocial issues in pediatric oncology. Med Pediatr Oncol. 1999;32(1): 44–8. Epub 1999/01/26.
40. Jankovic M, Spinetta JJ, Masera G, Barr RD, D'Angio GJ, Epelman C, et al. Communicating with the dying child: an invitation to listening – a report of the SIOP working committee on psychosocial issues in pediatric oncology. Pediatr Blood Cancer. 2008;50(5):1087–8. Epub 2008/03/25.

41. Hechler T, Blankenburg M, Friedrichsdorf SJ, Garske D, Hubner B, Menke A, et al. Parents' perspective on symptoms, quality of life, characteristics of death and end-of-life decisions for children dying from cancer. Klin Padiatr. 2008;220(3):166–74. Epub 2008/05/15.
42. Pettle SA, Britten CM. Talking with children about death and dying. Child Care Health Dev. 1995;21(6):395–404. Epub 1995/11/01.
43. McCabe MA. Involving children and adolescents in medical decision making: developmental and clinical considerations. J Pediatr Psychol. 1996;21(4):505–16. Epub 1996/08/01.
44. Kreicbergs U, Valdimarsdottir U, Onelov E, Henter JI, Steineck G. Talking about death with children who have severe malignant disease. N Engl J Med. 2004;351(12):1175–86. Epub 2004/09/17.
45. Freyer DR. Care of the dying adolescent: special considerations. Pediatrics. 2004;113(2): 381–8. Epub 2004/02/03.
46. Bell CJ, Skiles J, Pradhan K, Champion VL. End-of-life experiences in adolescents dying with cancer. Support Care Cancer. 2010;18(7):827–35. Epub 2009/09/04.
47. Levetown M. Communicating with children and families: from everyday interactions to skill in conveying distressing information. Pediatrics. 2008;121(5):e1441–60. Epub 2008/05/03.
48. McAliley LG, Hudson-Barr DC, Gunning RS, Rowbottom LA. The use of advance directives with adolescents. Pediatr Nurs. 2000;26(5):471–80. Epub 2002/05/25.
49. Lyon ME, Garvie PA, Briggs L, He J, McCarter R, D'Angelo LJ. Development, feasibility, and acceptability of the Family/Adolescent-Centered (FACE) Advance Care Planning intervention for adolescents with HIV. J Palliat Med. 2009;12(4):363–72. Epub 2009/03/31.
50. Wiener L, Ballard E, Brennan T, Battles H, Martinez P, Pao M. How I wish to be remembered: the use of an advance care planning document in adolescent and young adult populations. J Palliat Med. 2008;11(10):1309–13. Epub 2009/01/01.
51. Lyon ME, McCabe MA, Patel KM, D'Angelo LJ. What do adolescents want? An exploratory study regarding end-of-life decision-making. J Adolesc Health. 2004;35(6):529.e1–6. Epub 2004/12/08.
52. Gillam L, Sullivan J. Ethics at the end of life: who should make decisions about treatment limitation for young children with life-threatening or life-limiting conditions? J Paediatr Child Health. 2011;47(9):594–8. Epub 2011/09/29.
53. Brosig CL, Pierucci RL, Kupst MJ, Leuthner SR. Infant end-of-life care: the parents' perspective. J Perinatol. 2007;27(8):510–6. Epub 2007/04/20.
54. Tobler D, Greutmann M, Colman JM, Greutmann-Yantiri M, Librach SL, Kovacs AH. Knowledge of and preference for advance care planning by adults with congenital heart disease. Am J Cardiol. 2012;109(12):1797–800. Epub 2012/03/31.
55. Tobler D, Greutmann M, Colman JM, Greutmann-Yantiri M, Librach LS, Kovacs AH. End-of-life in adults with congenital heart disease: a call for early communication. Int J Cardiol. 2012;155(3):383–7. Epub 2010/11/26.
56. Reid GJ, Webb GD, Barzel M, McCrindle BW, Irvine MJ, Siu SC. Estimates of life expectancy by adolescents and young adults with congenital heart disease. J Am Coll Cardiol. 2006;48(2):349–55. Epub 2006/07/18.
57. Tobler D, Greutmann M, Colman JM, Greutmann-Yantiri M, Librach LS, Kovacs AH. End-of-life care in hospitalized adults with complex congenital heart disease: care delayed, care denied. Palliat Med. 2012;26(1):72–9. Epub 2011/06/24.
58. Greutmann M, Tobler D, Colman JM, Greutmann-Yantiri M, Librach SL, Kovacs AH. Facilitators of and barriers to advance care planning in adult congenital heart disease. Congenit Heart Dis. 2013;8(4):281–8. Epub 2013/01/03.
59. Kovacs AH, Landzberg MJ, Goodlin SJ. Advance care planning and end-of-life management of adult patients with congenital heart disease. World J Pediatr Congen Heart Surg. 2013;4(1): 62–9. Epub 2013/06/27.
60. Harrison JL, Silversides CK, Oechslin EN, Kovacs AH. Healthcare needs of adults with congenital heart disease: study of the patient perspective. J Cardiovasc Nurs. 2011;26(6):497–503. Epub 2011/03/05.

61. McCabe MA, Rushton CH, Glover J, Murray MG, Leikin S. Implications of the Patient Self-Determination Act: guidelines for involving adolescents in medical decision making. J Adolesc Health. 1996;19(5):319–24. Epub 1996/11/01.
62. Hsiao JL, Evan EE, Zeltzer LK. Parent and child perspectives on physician communication in pediatric palliative care. Palliat Support Care. 2007;5(4):355–65. Epub 2007/11/30.
63. Marino BS, Lipkin PH, Newburger JW, Peacock G, Gerdes M, Gaynor JW, et al. Neurodevelopmental outcomes in children with congenital heart disease: evaluation and management: a scientific statement from the American Heart Association. Circulation. 2012;126(9):1143–72. Epub 2012/08/02.
64. Lampert R, Hayes DL, Annas GJ, Farley MA, Goldstein NE, Hamilton RM, et al. HRS Expert Consensus Statement on the Management of Cardiovascular Implantable Electronic Devices (CIEDs) in patients nearing end of life or requesting withdrawal of therapy. Heart Rhythm. 2010;7(7):1008–26. Epub 2010/05/18.
65. Mitar M, Alba AC, MacIver J, Ross H. Lost in translation: examining patient and physician perceptions of implantable cardioverter-defibrillator deactivation discussions. Circ Heart Fail. 2012;5(5):660–6. Epub 2012/09/20.
66. Goodlin SJ, Quill TE, Arnold RM. Communication and decision-making about prognosis in heart failure care. J Card Fail. 2008;14(2):106–13. Epub 2008/03/08.
67. Morell E, Wolfe J, Scheurer M, Thiagarajan R, Morin C, Beke DM, et al. Patterns of care at end of life in children with advanced heart disease. Arch Pediatr Adolesc Med. 2012;166(8): 745–8. Epub 2012/04/05.
68. Donohoe RT, Haefeli K, Moore F. Public perceptions and experiences of myocardial infarction, cardiac arrest and CPR in London. Resuscitation. 2006;71(1):70–9. Epub 2006/09/02.
69. Groarke J, Gallagher J, McGovern R. Conflicting perspectives compromising discussions on cardiopulmonary resuscitation. Ir Med J. 2010;103(8):233–5. Epub 2010/11/05.
70. Keele L, Keenan HT, Sheetz J, Bratton SL. Differences in characteristics of dying children who receive and do not receive palliative care. Pediatrics. 2013;132(1):72–8. Epub 2013/06/12.

Chapter 10
Spiritual and Existential Issues

Stephanie Hooker and David B. Bekelman

Abstract
Background: Spiritual and existential issues become increasingly salient for individuals facing life-threatening illnesses, including those with advanced cardiovascular diseases. Patients with high levels of existential anxiety and low spiritual well-being tend to experience lower quality of life and more depressive symptoms. Thus, identifying patients who are having spiritual and existential issues and utilizing targeted spiritually-informed interventions may improve quality of care and quality of life at the end of life.

Method: We review the research examining spiritual and existential issues in patients with advanced cardiovascular disease at the end of life and provide recommendations for spiritual assessments and interventions providers can use with patients at the end of life.

Results: In general, observational studies indicate that patients with greater levels of spiritual well-being, religiousness, and meaning or purpose in life experience greater quality of life and fewer depressive symptoms. In contrast, negative aspects of spirituality or religiousness (e.g., existential anxiety) are related to poorer outcomes. Several spiritual assessments for patients at the end of life have been developed and can be used as guidelines for starting conversations with patients. However, few spiritually-integrated interventions have been tested and more research is needed to identify effective spiritual interventions.

Conclusions: Spiritual and existential issues are highly prevalent in patients at the end of life. Integrating spiritual care in advanced illness and at end of life can enhance the quality of care and patients' quality of life.

Keywords Spiritual • Existential • Palliative care • End of life • Cardiovascular diseases

S. Hooker, MS (✉)
Department of Psychology, University of Colorado Denver,
Campus Box 173, PO Box 173364, Denver, CO 80220–3364, USA
e-mail: stephanie.hooker@ucdenver.edu

D.B. Bekelman, MD, MPH
Department of Internal Medicine, VA Eastern Colorado Healthcare System and the University of Colorado School of Medicine, Denver, CO, USA

S.J. Goodlin, M.W. Rich (eds.), *End-of-Life Care in Cardiovascular Disease*,
DOI 10.1007/978-1-4471-6521-7_10

Key Points

- Spirituality and religiousness are important in the lives of many Americans, and spiritual and existential issues become more salient as individuals face life-threatening illnesses and death.
- Patients with greater spiritual well-being, meaning or purpose in life, and religious involvement tend to experience greater quality of life and fewer depressive symptoms.
- Providers can use spiritual assessments to identify patients experiencing spiritual or existential distress.
- Spiritually-informed interventions can enhance the quality of care at the end of life. However, few interventions have been formally studied, and therefore more research is needed to find effective interventions that integrate spiritual care into health care.

Introduction

> "I think [spirituality] is what heals you most. Medicine will only get you so far."
> 71-year-old female patient with chronic heart failure

"Spirituality," "religion," and "existential" are terms that are intuitively known but require definition. These terms have personal meanings to individuals, but all seem to have the core principle of "the search for the sacred" (i.e., a person, object, principle, or concept that transcends the self) [1]. Although spirituality and religiousness are often used interchangeably in the literature, they have notable definitional differences. Religion involves an identifiable group, prescribed behavioral patterns and religious expression (e.g., prayer, rituals), and may include the search for non-sacred goals (e.g., social support, a sense of belonging) [1]. Spirituality, on the other hand, does not usually involve an identifiable group but rather a sense of connection to something greater than the self [2] and the ability to "view life from a larger, more objective perspective" ([3], p. 988). Similar to spirituality, existential beliefs have been defined as relating to human existence and experience and what it means to be human; it involves a person's search for ultimate meaning, connections to others, or creative expression [4]. However, not all aspects of spirituality are positive. Existential anxiety has been defined as the concern that one's life has no meaning or purpose, and is considered to be the ultimate stressor [5].

Spirituality and religion are important aspects of the lives of many Americans, as a recent Gallup poll indicated that more than 90 % of Americans stated that they believed in God [6], and 59 % of the US population considers religion very or extremely important in daily life [7]. In recent years, both accrediting bodies and clinical practice guidelines have recognized spirituality as an important component of patient care. The Joint Commission on Accreditation for Health Care Organizations requires a spiritual assessment of patients in a variety of settings, including hospitals, home care organizations, long-term care facilities, and some behavioral health care organizations [8]. Additionally, both the World Health Organization [9] and the Clinical Practice Guidelines for Quality Palliative Care [10] highlight the importance of spiritual care in palliative care and emphasize the *integration* of psychological and spiritual aspects of patient care. Patients also believe that spirituality is important to integrate into their treatment [11] and medical schools are integrating coursework on spirituality into their curricula [12].

As patients face life-threatening illnesses and end of life, personal life meaning and spirituality may become more salient [13–15]. In a group of medically ill patients with a variety of acute and chronic health problems, all of the patients indicated that their spirituality or religious beliefs impacted their daily lives in various behavioral, affective, and somatic dimensions [16]. Thus, patients with advanced cardiovascular disease will also likely have spiritual beliefs and religious involvement. In one sample of heart failure patients, 98 % indicated that they had a religious affiliation [17] and in another study, a significantly greater proportion of heart failure patients (68.4 %) viewed religion as important compared to cancer patients [18]. Spirituality and religion are important aspects in the lives of patients with medical illnesses.

This chapter will provide a selective review of spiritual and existential issues in patients with advanced cardiovascular disease at the end of life. We have organized the chapter into three major sections, including observational studies, spiritual assessments, and spiritually informed interventions. Given the current state of the literature, we are limited to describing mostly observational studies as few spiritual interventions have been conducted. The spiritual assessments described in the chapter are relevant for all patients, not just those with advanced cardiovascular disease.

Observational Studies

Understanding which dimensions of spirituality and religion are related to health and quality of life can help clinicians and researchers fully understand *how* spirituality and religion influence health and well-being as well as identify targets for interventions. Therefore, the following review of observational findings is organized by outcome with studies that describe what aspects of spirituality or religiousness relate to each outcome.

Mortality and Clinical Outcomes

Several studies have examined the influence of religious or spiritual factors on mortality. In a meta-analysis of large studies examining the relationships between religious and spiritual factors and mortality, religious and spiritual factors were related to reduced risk for all-cause and cardiovascular-specific mortality in initially healthy samples but not in already diseased samples [19]. Conversely, very few studies have examined the relationship between religious or spiritual factors and cardiovascular morbidity, especially at the end of life. In two studies, spiritual or religious factors (e.g., frequency of religious participation, frequency of prayer or meditation, daily spiritual experiences) were not related to cardiovascular morbidity for initial [20] or recurrent [21] coronary events. In contrast to much of the literature, King et al. [22] found that *greater* strength of spiritual beliefs during an admission to the cardiology unit was related to *poorer* physician-rated clinical outcomes (i.e., improved or completely well, no change, or worse) at 9 months. There is some evidence that religious and spiritual factors are related to greater treatment adherence. Park and colleagues [23] found that greater religious commitment was related to greater treatment adherence in heart failure patients over the course of 6 months. In conclusion, religious and spiritual variables are largely protective of later mortality and cardiovascular morbidity in initially healthy patients. Although initial work suggests a positive relationship between religiousness and treatment adherence, more work needs to be done to understand how these factors influence clinical outcomes in diseased populations.

Quality of Life

The vast majority of research regarding spirituality at the end of life in cardiovascular disease has been conducted in relation to quality of life outcomes. Spiritual well-being, commonly related to quality of life, is defined as a sense of peace, compassion for others, a sense of reverence in life, and appreciation of both unity and diversity [24]. Spiritual well-being is often separated into two distinct dimensions: religious well-being and existential well-being. Religious well-being is the individual's perspective of his or her well-being in relation to a higher power whereas existential well-being is his or her perspective of intrapersonal and interpersonal concerns of the self, others, and the environment. Existential well-being involves feeling a sense of purpose in life and interconnectedness to others and the world. One study directly compared the spiritual well-being of symptomatic heart failure patients to advanced cancer patients and found that they were not significantly different [25]. In a study of heart failure outpatients, existential and religious well-being were both positively related to general well-being and negatively related to both physical and emotional symptoms [26]. Together, existential and religious well-being accounted for 24 % of the variance in general well-being. Similarly, Bekelman and colleagues [27] found that heart failure patients with a greater sense of meaning/peace (similar to existential well-being) reported less depressive symptoms and greater heart failure-specific quality of life. Faith (similar to religious well-being) was less strongly associated with depressive symptoms and quality of life. Heart failure patients with greater spiritual well-being also reported greater perceived control over their heart conditions, with existential well-being being more strongly related to perceived control than religious well-being [28]. These patterns of findings suggest that existential well-being, or having a greater sense of meaning and purpose, is more important for quality of life than religious well-being or faith.

Meaning in life, or the extent to which one feels that existence is significant and activities in life are valuable [29], is the core component of existential well-being. Identifying what gives meaning to the lives of patients with heart failure could generate targets for intervention to increase meaning in life. In a qualitative study of heart failure patients, the majority of patients (54 %) stated that their families gave their lives meaning in the context of the illness, followed by helping others or working (13 %), God, religion, faith, or morality (11 %), and hobbies or leisure activities (9 %) [30]. Also in heart failure patients, Park and colleagues [31] found that meaning in life was positively related to both mental and physical health-related quality of life. However, on average, meaning in life decreased over the course of 6 months. For a subset of patients who used their faith to cope with the illness and accepted and tried to understand their illness in a more positive light, meaning in life increased over the course of the 6 months. Therefore, meaning may be an important target for interventions in patients with cardiovascular disease at the end of life.

Intrinsic religiousness is a motivation to engage in religion because of an individual's religious beliefs and is often contrasted with extrinsic religious motivation, which is the motivation to engage in religion because of social factors or emotional comfort [32]. In one study of chronic cardiac patients (i.e., previous myocardial infarction, ischemia, angina pectoris, heart failure, arrhythmias), intrinsic religiousness was negatively associated with helplessness and positively associated with illness acceptance, perceived benefits of having an illness, physical status, and emotional well-being after controlling for gender, age, and educational level [33]. Thus, intrinsic religiousness may help patients cope with their illness and therefore reduce distress, although more research is needed to test this hypothesis.

Few studies have examined changes in spirituality over time. Park [34] asked NYHA III or IV heart failure patients to estimate how long they had to live ("How long do you think you are going to live?") on a scale from 1 (*I don't know*) to 8 (*more than 10 years*) twice, once at baseline and once 6 months later. She found that decreasing estimates of longevity over the 6 months were related to increases in forgiveness and perceptions of a meaningful life during the same time period. Moreover, decreasing estimates of longevity were marginally related to less spiritual struggle. This suggested that patients who came to terms with the idea that their lives were coming to an end also experienced enhanced spiritual well-being, although a causal relationship could not be determined from this study. Patients who experienced enhanced spiritual well-being may have also been more likely to accept the uncertainty of death and to resolve spiritual issues.

Spiritual needs, or the human need to find meaning, purpose, and value in one's life [35], also influence quality of life at the end of life. Murray and colleagues [35] interviewed patients with NYHA grade IV heart failure and patients with inoperable lung cancer regarding their experiences and spiritual beliefs up to four times (every 3 months) in a year. They found that heart failure patients often struggled to find meaning in life and those who found comfort in spiritual or religious beliefs reported increased feelings of strength and hope. These results suggested that spiritual care may benefit heart failure patients with high spiritual distress.

Depression

Depression is highly prevalent in heart failure patients [36, 37], and some spiritual and religious factors may be protective against depressive symptoms at the end of life. Greater church attendance was related to better physical health (i.e., self-reported illness severity and physician-rated NYHA class), less depression, and greater social support in heart failure patients [17]. Compared to depressed heart failure patients, heart failure patients who were not depressed were more likely to rate themselves as both religious and spiritual, to attend religious services, and to engage in daily spiritual activities (e.g., reading the Bible and daily prayer). Patients with no religious affiliation were 12.7 and 10.3 times more likely to have major and minor depression, respectively, than patients with a religious affiliation [38]. There is also evidence that meaning in life is significantly related to less depression and physical symptoms. The relationship between depression and symptom burden in heart failure patients was significantly moderated by purpose in life and self-transcendence, in that patients with greater meaning in life experienced less symptom burden [39].

End of Life Decision-Making

In addition to evidence that spirituality is related to quality of life, there is also evidence that spirituality is related to decision-making at end of life. For example, patients who grew closer to God as a result of their illnesses or grew spiritually as a result of their illnesses were more likely to accept the risk associated with a high burden life-sustaining treatment. The authors suggested that patients who grew spiritually might have had greater hope in the likelihood that the outcome will be favorable and if it was not favorable, that they could endure the burden with the aid of spiritual resources [40].

Traditionally, religious involvement (e.g., religious affiliation, religious service attendance) and daily spiritual experiences (e.g., praying, reading the Bible) have received the greatest attention in the religion and health literature. In a study of patients undergoing coronary artery bypass graft (CABG) surgery, there were mixed associations between spiritual and religious factors and whether or not they engaged in end of life decision-making ("Are your affairs in order in case you die?") [41]. After controlling for age, surgery complexity, comorbidities, and functional status, patients who engaged in more private religious practices (e.g., private prayer, religious television programs, reading religious books) and reported greater reverence in secular contexts (e.g., in nature, enjoying music or art) were more likely to engage in end of life decision-making prior to CABG. Conversely, those who reported more reverence in religious contexts (e.g., attending services, reading the Bible) and who engaged in petitionary prayer were less likely to engage in end of life decision-making prior to CABG. The authors suggested that a possible explanation for their mixed findings was that those who had greater intrinsic religiosity (and therefore engaged in private religious practices and had a strong sense of secular reverence) would likely have had less death anxiety and therefore were more confident in their end of life decisions. However, future research is needed to test this hypothesis.

Not All Spiritual Influences Are Positive

Westlake and colleagues [42] noted in their review that patients with heart failure often experienced spiritual or existential suffering and reported that their lives were lacking meaning and purpose. Patients with NYHA III or IV heart failure who reported greater religious struggle, or the belief that God is punishing or abandoning one in times of stressful life circumstances, also reported more concerns about death and depressive symptoms [43]. Although the vast majority of evidence regarding the relationships between spirituality and quality of life or emotional well-being outcomes in patients with cardiovascular disease is salutary, there are a few examples of studies with null findings. Religious coping, or using religious or spiritual resources to manage stress and anxiety, was not significantly related to depressed affect in patients with mild to moderate heart failure prospectively at 6 months [44].

Spiritual Assessment

"Diagnosing" spiritual distress in patients at the end of life proves difficult, as the spiritual experience can be very diverse. For example, Oates [13] noted that spiritual distress could take on a form of a search for meaning, seeking forgiveness, or expressing needs for love and hope. Murray and colleagues [35], on the other hand, reported that many patients did not explicitly express spiritual needs, but expressed feeling "low" and feeling their lives were valueless or useless. Williams [45] suggested that the spiritual phenomena that occur at the end of life are spiritual despair, spiritual work, and spiritual well-being. Similarly, Westlake and Dracup [30] suggested that there is a three-step process through which spirituality contributes to adjustment to heart failure: (1) regret regarding past behaviors and lifestyles; (2) the search for meaning in the heart failure; and (3) the search for hope in the future. Identifying patients who are in spiritual despair, are regretting their past lifestyle choices, or are searching for meaning in the context of the illness are important goals of spiritual assessment.

Spiritual assessment is a process of gathering and synthesizing information about a patient's spiritual and religious beliefs in order to make more informed decisions about that patient's care [8]. "Paper and pencil" questionnaires used in research can provide initial information regarding spirituality, but we and others believe they should not be used as the only method of assessment [46]. We recommend using open-ended questions and careful listening while having a conversation with patients to understand the role of spirituality or religion in their lives. For example, asking, "What role does spirituality or religion play in your life" can be a useful opening, with follow up probing questions (e.g., "tell me more about that") based on the patient's response [47]. Although a detailed account of spiritual assessment is beyond the scope of the chapter, we provide a brief overview of the assessments in Table 10.1 and discuss a few measures in depth.

Spiritual needs in advanced illness have been classified into several domains, and exploring each of these domains can serve as a framework for assessment. Groves and Klauser [48] describe four domains of spiritual pain, including (1) meaning (e.g., the inability to make sense of what is happening), (2) forgiveness (e.g., the failure to forgive oneself or others), (3) relatedness (e.g., the loss of relationships and roles) and (4) hopelessness. Sulmasy adds the domain of value, the dignity people have because they are human beings, which is challenged during serious illness because of changes in appearance and abilities. Sulmasy provides a list of questions that can be used to inquire into issues of value, relationship, and meaning [47]. A single question, "Are you at peace?" can be used to begin a discussion of spiritual issues [49]. This question was compared to general quality of life and spiritual well-being in a mixed sample of patients at the end of life (cancer, heart failure, end-stage renal disease, chronic obstructive pulmonary disease). The one-item measure was strongly and positively correlated with emotional and spiritual well-being and moderately and positively correlated with physical well-being, functional well-being, and social well-being.

Several guidelines for conducting structured spiritual assessments have been developed, including HOPE [50], FICA [51], and SPIRIT [52]. In addition, the Working Group on Religious and Spiritual Issues at the End of Life Guidelines for Physicians provides a list of conversation goals and example phrases to help elicit patient concerns [53]. These systems have several commonalities, including identifying how important religion or spirituality is for the patient, how they believe it affects their illness or medical decision-making, and whether the patient has any current spiritual concerns. These assessments provide an outline to ensure providers touch on important aspects of religiousness or spirituality, and can be used as guides for starting conversations with their patients about religious or spiritual needs.

Spiritual Interventions

Spiritual care has been defined as supporting another person in spiritual distress by being present, deeply listening, and being compassionate [54]. Clergy or chaplains have been the primary providers of spiritual care and are trained to address spiritual and existential concerns [54]. However, studies show that many patients want

Table 10.1 Spiritual assessments

Author	Instrument	Number of items or domains	Description
Sulmasy [47]		1	"What role does spirituality or religion play in your life?"
Cobb [66]		1	"I see that you describe your religion as ______. Can you tell me more about this?"
Steinhauser et al. [49]		1	"Are you at peace?"
Anandarajah and Hight [50]	HOPE	4	H: Assessment of sources of hope, strength, comfort, meaning, peace, love, and connection.
			O: Assessment of organized religion.
			P: Assessment of personal spirituality and practices.
			E: Effects of spirituality on care and end-of-life decisions.
Puchalski and Romer [51]	FICA	4	F: Faith or belief: what is your faith or belief?
			I: Importance: is it important in your life?
			C: Community: are you part of a religious or spiritual community?
			A: Address: How would you like me, your health-care provider, to address these issues in your health care?
Maugens [52]	SPIRIT	6	S: Spiritual belief system: what is your religious affiliation?
			P: Personal Spirituality: What is the importance of spirituality in your daily life?
			I: Integration with a spiritual community: Do you belong to any spiritual or religious group?
			R: Ritualized practices and restrictions: Are there aspects of medical care that you forbid on religious/spiritual grounds?
			I: Implications for medical care: What aspects of your religion/spirituality would you like me to keep in mind as I care for you?
			T: Terminal events planning: As we plan for your care near the end of life, how does your faith impact your decisions?

healthcare providers, including physicians, to consider their spiritual or religious beliefs or needs [11, 55, 56], yet 68 % of patients reported that no physician had ever inquired about their spiritual or religious needs [11]. Thus, there is room for healthcare providers to integrate spirituality in the care of patients at the end of life. The role of healthcare providers is to facilitate patients' experience of spirituality and religiousness, not to give patients' meaning or to "treat" their spiritual distress [47]. We provide several practical applications for addressing spiritual distress (see Table 10.2 for an overview) and describe the current state of the research in spiritually informed interventions in patients with advanced cardiovascular disease.

Frankl [57] suggested four areas in which reflection may encourage patients to reconnect to their sense of meaning. These areas included: (1) their creations or accomplishments; (2) who or what have they loved; (3) what legacy have they left behind; and (4) the things they believe in and how can they transcend the suffering. Spira [58] described several existential exercises that may help alleviate existential anxiety. For example, he suggested having patients consider what their optimal futures would be ("If you had 1 year (month, week) to live and you wanted to make it the most meaningful year of your life, what would you do?") and reprioritizing life activities so that patients are living in accordance with their values. Although these brief interventions have not been formally tested, they provide a basic way for providers to help patients reconnect to what made their lives meaningful and perhaps reduce some of the existential anxiety they are experiencing.

There are many other practices that may be helpful for spiritual distress besides talking. Groves and Klauser [48] describes a number of them, including art and music therapies, guided visualization, meditation, and various rituals (see Table 10.2 for a more complete list). Sometimes a ritual or behavior is a good alternative to talking. Consider for example the importance of birth, death, and marital rituals in various cultures.

Conducting interventional studies in end of life populations can be challenging giving attrition from clinical trials. In one recent trial in patients with advanced solid tumors, only 43.1 % (66 of 153) participants completed a 2-month study protocol [59]. To our knowledge, only two interventions (both in the pilot phase) incorporating spirituality have been formally studied in patients with cardiovascular disease at the end of life. Delaney et al. [60] pilot tested a spirituality-based intervention for community-dwelling patients with cardiovascular disease and Masters and colleagues [61] pilot tested a 12-week mail-based psychospiritual intervention for patients with heart failure. Both interventions showed promising results, as they were deemed feasible to implement and showed significant increases in quality of life. However, future studies are needed to further test the efficacy of these interventions.

There is much room for the research to grow in this area. Spiritually informed interventions for patients with cancer, such as meaning-centered group therapies [62] and dignity therapy [63], are interventions that have had some success and could potentially be adapted for patients with advanced cardiovascular disease. Breitbart and colleagues [62] piloted a meaning-centered group psychotherapy intervention for patients with advanced cancer. Influenced by Frankl's work, eight

Table 10.2 Spiritual and existential interventions

Author	Approach
Frankl [57]	Have patients reflect on four areas
	1. What are your creations or accomplishments?
	2. Who or what have you loved?
	3. What legacy have you left behind?
	4. What are the things you believe in? How can you use these beliefs to cope with the suffering?
Spira [58]	*Considering an Optimal Future:* "If you had 1 year (month, week) to live and you wanted to make it the most meaningful year of your life:
	What personal characteristics (self-image, personality, assumptions about life) would you want to let go of?
	What personal characteristics would you want to have to help you make this year a valuable one for you?
	What activities would you want to engage in that would bring greatest meaning and value to your life?
	What is stopping you from having these qualities and doing these activities now?
	What can you do to overcome these barriers and live more fully?"
	Reprioritizing Life Activities: Helps patients identify meaningful actions and prioritize spending more time in activities that are meaningful to them.
	1. Make a list of activities you spend your time doing in a typical week.
	2. Prioritize this list, with the activities you spend most time with at the top and those you spend less time on toward the bottom.
	3. Prioritize these activities again, but this time list at the top the activities that bring more meaning and personal value to your life, with progressively less meaningful activities toward the bottom. Any activities that you wish you were doing, even though you have not gotten around to them, can also be added to this list.
	4. Examine the last two columns. If the lists are ordered differently, then it is important to ask what you can do to spend more time engaged in those activities that bring more meaning and value to your life and less in activities that you do out of habit.
Groves and Klauser [48]	Art therapy
	Breath work
	Dream work
	Energy therapies
	Forgiveness exercises
	Guided visualization
	Prayer
	Journaling
	Life review exercises
	Meditation
	Music therapy
	Religious rites and sacred writings
	Rituals for the bedside
	Rituals of remembering

group sessions explored concepts and sources of meaning, including the impact of cancer on one's sense of meaning and identity. Preliminary evidence suggested that compared to a supportive group therapy control, patients in the meaning-centered group experienced greater improvements in spiritual well-being and a sense of meaning as well as decreases in anxiety and desire for hastened death.

Dignity therapy involves the creation of a legacy document [63] via a brief individual psychotherapy designed to address issues of generativity, meaning, and purpose at the end-of-life by having patients reflect on the things that have mattered most to them in life and focus on how they would like to be remembered. Compared to a client-centered psychotherapy and a standard palliative care control, cancer patients who received dignity therapy were more likely to report that the intervention helped them by improving their quality of life and sense of dignity. However, there were no significant differences in depressive or anxiety symptoms, quality of life, spiritual well-being, or physical symptoms between groups or between baseline and follow-up assessments. The researchers concluded that because patients were not very distressed at baseline, their ability to find significant improvements was limited. Chochinov [63] lists the questions used in dignity therapy, which may useful in advanced cardiovascular disease. Further research testing interventions similar to those in oncology and novel interventions that address spirituality at the end of life in patients with cardiovascular disease is needed.

Conclusions

Spirituality and existential issues are highly prevalent in patients with cardiovascular disease at the end of life [16–18]. In general, positive aspects of spirituality (e.g., meaning in life, spiritual well-being, religious or spiritual involvement) are positively related to health and well-being whereas as negative aspects (e.g., religious or spiritual struggle, existential anxiety) are negatively related to health and well-being. This is consistent with findings in healthy samples and in patients with other chronic illnesses [64]. Therefore, identifying patients with significant spiritual struggles or existential anxiety is crucial to providing quality care to patients at the end of life.

There are several methods to assess spirituality in patients, yet many authors simply recommend having a conversation with patients about spirituality. Several guidelines are available to pick domains and questions that may be relevant in the assessment. There are many approaches to address spiritual distress, although very few have been empirically studied.

Research in this area would benefit from consensus in the definitions of religiousness/spirituality and how to measure each of the constructs. Spirituality and religiousness are often measured in several different ways, making it difficult to draw conclusions about which components of spirituality are truly beneficial for health. Miller and Thoresen [65] further noted the difficulties in definitions and measurement stating that scientific definitions of religiousness and spirituality are likely to differ from those of the layperson. Thus, multidimensional measures of spirituality should be used when possible, and spirituality should be considered within the cultural context of the participants as a whole [64].

Considering the importance of integrating spirituality in palliative care [9, 10], future research is needed to identify effective intervention strategies. The two interventions currently in the pilot phase with patients with cardiovascular disease are promising; yet efficacy studies and further intervention development are needed.

References

1. Hill PC, Pargament KI, Hood RW, et al. Conceptualizing religion and spirituality: points of commonality, points of departure. J Theory Soc Behav. 2000;30:51–77.
2. Hyland ME, Wheeler P, Kamble S, Masters KS. A sense of 'special connection', self-transcendent values and a common factor for religious and non-religious spirituality. Arch Psychol Relig. 2010;32:292–326.
3. Piedmont RL. Does spirituality represent the sixth factor of personality? Spiritual transcendence and the five-factor model. J Pers. 1999;67:985–1103.
4. Selman L, Beynon T, Higginson IJ, Harding R. Psychological, social and spiritual distress at the end of life in heart failure patients. Curr Opin Support Palliat Care. 2007;1:260–6.
5. Morse DR. Confronting existential anxiety: the ultimate stressor. Stress Med. 1998;14:109–19.
6. Newport F. More than 9 in 10 Americans continue to believe in God. Gallup. 3 June 2011. Available at http://gallup.com/poll/147887/Americans-Continue-Believe-God.aspx. Accessed 27 Jul 2013.
7. Religion in America [news release]. New York: CBS News. 13 Apr 2006. http://www.cbsnews.com/htdocs/CBSNews_polls/religion_041306.pdf.
8. Hodge DR. A template for spiritual assessment: a review of the JCAHO requirements and guidelines for implementation. Soc Work. 2006;51:317–26.
9. World Health Organization Definition of Palliative Care. Available at: http://www.who.int/cancer/palliative/definition/en/. Accessed 8 Aug 2013.
10. National Consensus Project for Quality Palliative Care. Clinical practice guidelines for quality palliative care, 2nd ed; 2009. Available at http://www.nationalconsensusproject.org. Accessed 27 Jul 2013.
11. King DE, Bushwick B. Beliefs and attitudes of hospital inpatients about faith healing and prayer. J Fam Pract. 1994;39:349–52.
12. Puchalski CM. Spirituality and medicine: curricula in medical education. J Cancer Educ. 2006;21:14–8.
13. Oates L. Providing spiritual care in end-stage cardiac failure. Int J Palliat Nurs. 2004;10:485–90.
14. Pulchaski CM, Dorff RE, Hendi IY. Spirituality, religion, and healing in palliative care. Clin Geriatr Med. 2004;20:689–714.
15. Sulmasy DP. A biopsychosocial-spiritual model for the care of patients at the end of life. Gerontologist. 2002;42(Special Issue III):24–33.
16. Woods TE, Ironson GH. Religion and spirituality in the face of illness: how cancer, cardiac, and HIV patients describe their spirituality/religiosity. J Health Psychol. 1999;4:393–412.
17. Koenig HG. Religion, congestive heart failure, and chronic pulmonary disease. J Relig Health. 2002;41:263–78.
18. Kub JE, Nolan MT, Hughes MT, et al. Religious importance and practices of patients with a life-threatening illness: implications for screening protocols. Appl Nurs Res. 2003;16:196–200.

19. Chida Y, Steptoe A, Powell LH. Religiosity/spirituality and mortality. Psychother Psychosom. 2009;78:81–90.
20. Feinstein M, Liu K, Ning H, Fitchett G, Lloyd-Jones DM. Burden of cardiovascular risk factors, subclinical atherosclerosis, and incident cardiovascular events across dimensions of religiosity: the multi-ethnic study of atherosclerosis. Circulation. 2010;121(5): 659–66.
21. Blumenthal JA, Babyak MA, Ironson G, et al. Spirituality, religion, and clinical outcomes in patients recovering from an acute myocardial infarction. Psychosom Med. 2007;69: 501–8.
22. King M, Speck P, Thomas A. The effect of spiritual beliefs on outcome from illness. Soc Sci Med. 1999;48:1291–9.
23. Park CL, Moehl B, Fenster JR, Suresh DP, Bliss D. Religiousness and treatment adherence in congestive heart failure patients. J Relig Spirituality Aging. 2008;20:249–66.
24. Paloutzian R, Ellison C. Spiritual well-being: conceptualization and measurement. J Psychol Theol. 1983;11:330–40.
25. Bekelman DB, Rumsfeld JS, Havranek EP, et al. Symptom burden, depression, and spiritual well-being: a comparison of heart failure and advanced cancer patients. J Gen Intern Med. 2009;24:592–8.
26. Beery TA, Baas LS, Fowler C, Allen G. Spirituality in persons with heart failure. J Holist Nurs. 2002;20:5–25.
27. Bekelman DB, Parry C, Curlin FA, Yamashita TE, Fairclough DL, Wamboldt FS. A comparison of two spirituality instruments and their relationship with depression and quality of life in chronic heart failure. J Pain Symptom Manage. 2010;39:515–26.
28. Vollman MW, LaMontagne LL, Wallston KA. Existential well-being predicts perceived control in adults with heart failure. Appl Nurs Res. 2009;22:198–203.
29. Reker GT, Peacock EJ, Wong PT. Meaning and purpose in life and well-being: a life-span perspective. J Gerontol. 1987;42:44–9.
30. Westlake C, Dracup K. Role of spirituality in adjustment of patients with advanced heart failure. Prog Cardiovasc Nurs. 2001;16:119–25.
31. Park CL, Malone MR, Suresh DP, Bliss D, Rosen RI. Coping, meaning in life, and quality of life in congestive heart failure patients. Qual Life Res. 2008;17:21–6.
32. Gorsuch RL, McPherson SE. Intrinsic/extrinsic measurement: I/E-revised and single-item scales. J Sci Stud Relig. 1989;28:348–54.
33. Karademas EC. Illness cognitions as a pathway between religiousness and subjective health in chronic cardiac patients. J Health Psychol. 2010;15:239–47.
34. Park CL. Estimated longevity and changes in spirituality in the context of advanced congestive heart failure. Palliat Support Care. 2008;6:3–11.
35. Murray SA, Kendall M, Boyd K, Worth A, Benton TF. Exploring the spiritual needs of people dying of lung cancer or heart failure: a prospective qualitative interview study of patients and their careers. Palliat Med. 2004;18:39–45.
36. Rumsfeld JS, Havranek E, Masoudi FA, et al. Depressive symptoms are the strongest predictors of short-term declines in health status in patients with heart failure. J Am Coll Cardiol. 2003;42(10):1811–7.
37. Rutledge T, Reis VA, Linke SE, Greenberg BH, Mills PJ. Depression in heart failure: a meta-analytic review of prevalence, intervention effects, and associations with clinical outcomes. J Am Coll Cardiol. 2006;48(8):1527–37.
38. Koenig HG. Religion and depression in older medical inpatients. Am J Geriatr Psychiatry. 2007;15:282–91.
39. Gusick GM. The contributions of depression and spirituality to symptom burden in heart failure. Arch Psychiatry Nurs. 2008;22:53–5.
40. Van Ness PH, Towle VR, O'Leary JR, Fried TR. Religion, risk, and medical decision making at the end of life. J Aging Health. 2008;20:545–59.
41. Ai AL, Park CL, Shearer M. Spiritual and religious involvement relate to end-of-life decision-making in patients undergoing coronary bypass graft surgery. Int J Psychiatry Med. 2008;38: 113–32.

42. Westlake C, Dyo M, Vollman M, Heywood JT. Spirituality and suffering of patients with heart failure. Prog Palliat Care. 2008;16:257–65.
43. Edmondson D, Park CL, Chaudoir SR, Wortmann JH. Death without God: religious struggle, death concerns, and depression in the terminally ill. Psychol Sci. 2008;19:754–8.
44. Park CL, Sacco SJ, Edmondson D. Expanding coping goodness-of-fit: religious coping, health locus of control, and depressed affect in heart failure patients. Anxiety Stress Coping. 2012;25:137–53.
45. Williams A. Perspectives on spirituality at the end of life: a meta-summary. Palliat Support Care. 2006;4:407–17.
46. Taylor EJ. Spiritual assessment. In: Ferrell BR, Coyle N, editors. Oxford textbook of palliative nursing. 3rd ed. New York: Oxford University Press; 2010. p. 647–61.
47. Sulmasy DP. Spiritual issues in the care of dying patients "…It's okay between me and God". JAMA. 2006;296(11):1385–92.
48. Groves RF, Klauser HA. The American book of dying: lessons in healing spiritual pain. Berkeley: Celestial Arts; 2005.
49. Steinhauser KE, Voils CI, Clipp EC, Bosworth HB, Christakis NA, Tulsky JA. Are you at peace? One item to probe spiritual concerns at the end of life. Arch Intern Med. 2006;166:101–5.
50. Anandarajah G, Hight E. Spirituality and medical practice: using the HOPE questions as a practical tool for spiritual assessment. Am Fam Physician. 2001;63:81–8.
51. Puchalski C, Romer AL. Taking a spiritual history allows clinicians to understand patients more fully. J Palliat Med. 2000;3:129–37.
52. Maugens TA. The spiritual history. Arch Fam Med. 1996;5:11–6.
53. Lo B, Ruston D, Kates LW, et al. Discussing religious and spiritual issues at then end of life: a practical guide for physicians. JAMA. 2002;287:749–54.
54. Baird P. Spiritual care interventions. In: Ferrell BR, Coyle N, editors. Oxford textbook of palliative nursing. 3rd ed. New York: Oxford University Press; 2010. p. 663–71.
55. Maugans TA, Wadland WC. Religion and family medicine: a survey of physicians and patients. J Fam Pract. 1991;32:210–3.
56. Ehman JW, Ott B, Short TH, Ciampa RC, Hansen-Flaschen J. Do patients want physicians to inquire about their spiritual or religious beliefs if they become gravely ill? Arch Intern Med. 1999;159:1803–6.
57. Frankl V. Man's search for meaning- an introduction to logotherapy. London: Hodder and Stoughton; 1987.
58. Spira JL. Existential psychotherapy in palliative care. In: Chochinov HM, Breitbart W, editors. Handbook of psychiatry in palliative medicine. New York: Oxford University Press; 2000. p. 197–214.
59. Appelbaum AJ, Lichtenthal WG, Pessin HA, et al. Factors associated with attrition from a randomized controlled trial of meaning-centered group psychotherapy for patients with advanced cancer. Psychol Oncol. 2012;21:1195–204.
60. Delaney C, Barrere C, Helming M. The influence of a spirituality-based intervention on quality of life, depression, and anxiety in community-dwelling adults with cardiovascular disease: a pilot study. J Holist Nurs. 2011;29:21–32.
61. Hooker SA, Ross K, Masters KS, et al. Qualitative evaluation of a psychospiritual intervention for chronic heart failure (heart failure) patients: the Denver Spirited Heart Pilot Study. Ann Behav Med. 2013;45 Suppl 2:s64.
62. Breitbart W, Rosenfeld B, Gibson C, et al. Meaning-centered group psychotherapy for patients with advanced cancer: a pilot randomized controlled trial. Psychol Oncol. 2010;19:21–8.
63. Chochinov HM, Kristjanson LJ, Breitbart W, McClement S, Hack T, Hassard T, Harlos M. Effect of dignity therapy on distress and end-of-life experience in terminally ill patients: a randomized controlled trial. Lancet Oncol. 2011;12:753–62.
64. Masters KS, Hooker SA. Religiousness/spirituality, cardiovascular disease, and cancer: cultural integration for health research and intervention. J Consult Clin Psychol. 2013;81:206–16.
65. Miller WR, Thoresen CE. Spirituality, religion, and health: an emerging research field. Am Psychol. 2003;58:24–35.
66. Cobb M. Spiritual care. In: Lloyd-Williams M, editor. Psychosocial issues in palliative care. New York: Oxford University Press; 2003. p. 135–47.

Chapter 11
Management of Implanted Cardiac Rhythm Devices at End of Life

Rachel Lampert

Abstract Implantable cardioverter-defibrillator (ICD) deactivation can improve quality of life in patients nearing the end of life, and all ICD patients should have the opportunity to discuss this option. Ethical principles and legal precedents support deactivation in patients who wish withdrawal of this life-sustaining therapy. Physicians must communicate this option to patients throughout their course, and particularly when advanced care planning or "Do Not Resuscitate" status is under discussion. While the logistics of deactivation can be challenging, inpatient and outpatient hospices should have policies in place for ICD deactivation.

Keywords Implantable cardioverter defibrillator • Deactivation • Shocks • Quality of life • Palliative care

Key Points

- ICD deactivation can improve quality of life in patients who are nearing the end of life.
- All ICD patients should have the opportunity to discuss this option.
- Ethical principles and legal precedents support deactivation in patients who wish withdrawal of this life-sustaining therapy.
- Physicians must communicate this option to patients throughout the course of illness, and particularly when advanced care planning or "DNR" status is under discussion.
- While the logistics of deactivation can be challenging, inpatient and outpatient hospices should have policies in place for deactivation.

R. Lampert, MD
Section of Cardiology, Yale University School of Medicine,
789 Howard Avenue, Dana 319, New Haven, CT 06520, USA
e-mail: Rachel.lampert@yale.edu

S.J. Goodlin, M.W. Rich (eds.), *End-of-Life Care in Cardiovascular Disease*,
DOI 10.1007/978-1-4471-6521-7_11

Case 1

An interview with a family-member of a recently-deceased patient with an implantable cardioverter-defibrillator (ICD):

"His defibrillator kept going off…It went off 12 times in one night…He went in and they looked at it…they said they adjusted it and they sent him back home. The next day we had to take him back because it was happening again. It kept going off and going off and it wouldn't stop going off." [1]

Case 2

A 65 year old man received an ICD 10 years ago following a cardiac arrest. The left ventricular ejection fraction (LVEF) was 30 %. He had shock-treated ventricular tachycardia at 250 bpm four times in the last 10 years. He was last seen 8 months ago with subsequent transtelephonic follow-up. His wife called his electrophysiologist: Mr. G was currently on the oncology floor, having been diagnosed with metastatic cancer 4 months ago. Despite multiple rounds of chemotherapy, he had continued lesions in brain and bone. He was to be discharged to hospice the following day on a morphine drip. The patient and family requested that the ICD be turned off. The electrophysiology team deactivated the ICD, and the patient died in hospice 2 days later, "peacefully" according to his wife [2].

Indications for ICDs continue to expand [3], with over 10,000 new devices implanted each month in the US [4]. While the ICD protects against sudden cardiac death, all individuals will eventually succumb to terminal illnesses. As the population ages, the numbers of patients who have received ICDs and are now dying of other causes, whether from cardiac disease or other terminal illnesses, will continue to grow. The pain and palliative care literature abounds with reports of patients receiving ICD shocks at the end of life, to the distress of patients and families, as illustrated in the first case above. Studies show that these are not rare events—10–20 % of patients receive shocks from their ICDs in the last days or weeks of their lives [1, 5]. Shocks have been described as a "blow to the chest" or "being kicked by a mule" [6] and thus it is not surprising that the pain, anxiety, and fear that occur with or in anticipation of these shocks can decrease quality of life [7, 8].

However, dying with an ICD in place does not have to be accompanied by painful shocks as illustrated by the second case above, in which the ICD was deactivated following communication amongst patients, families, and healthcare providers. Several recent consensus statements have suggested ways to ensure that patients with ICDs have the opportunity to discuss the option of deactivation, and have described the logistical aspects of deactivation [9, 10]. Deactivation prevents shocks as patients are dying [5]. However, among ICD patients who have died, few have had the option to discuss deactivation before death [1, 11, 12]. In a subanalysis of the MADIT II study, among 98 patients who died, just 15 had had their devices deactivated; [5] even among patients in hospice, less than half had their ICD deactivated [13].

There are several barriers to ICD deactivation that have prevented the widespread adoption of this practice necessary to improve quality of life as patients reach the end of life. These include varying perceptions of the legal and ethical issues, difficulties with communication between health care providers and patients, and logistical issues.

Legal and Ethical Issues

Most healthcare providers who care for patients with cardiac implantable electronic devices (CIEDs) have participated in device deactivations [14]. While clear-cut ethical and legal principles support deactivation, understanding of the issues surrounding device deactivation by healthcare providers varies [14, 15]. The first fundamental ethical and legal principle is informed consent, which is the most important legal doctrine in the clinician-patient relationship. Informed consent derives from the ethical principle of respect for persons: autonomy is maximized when patients understand the nature of their diagnoses and treatment options and participate in decisions about their care [16]. Clinicians are ethically and legally obligated to ensure that patients are informed about their diagnoses and treatment options [17, 18]. US courts have ruled that the right to make decisions about medical treatments is both a common law right (i.e. derived from court decisions) based on bodily integrity and self-determination and a constitutional right based on privacy and liberty [19].

A corollary to informed consent is informed refusal. A patient has the right to refuse any treatment, even if the treatment prolongs life or if death would follow treatment refusal. A patient also has the right to refuse a previously consented treatment if the treatment no longer meets the patient's health care goals because those goals have changed (e.g., from prolonging life to minimizing discomforts), or if the perceived burdens of the patient's illness and its treatment outweigh the perceived benefits of therapy. US courts have consistently upheld the right of a patient to withdraw life-sustaining therapies, as outlined in Table 11.1 [18, 19, 20–22]. In none of these cases did the courts distinguish between types of life-sustaining treatments. The law applies to the person, and informed consent is a right of the patient—it is not specific to any one medical intervention [19, 23–25]. Thus, even though the Supreme Court has not specifically commented on the question of pacemaker (PM) or ICD deactivation, because CIEDs deliver life-sustaining therapies, discontinuation of these therapies is clearly addressed by Supreme Court precedents upholding the right to discontinue life-sustaining treatment.

Further, while some healthcare providers feel that deactivation of CIEDs is akin to euthanasia [14], US Supreme Court decisions have made a clear distinction between withdrawing life-sustaining treatments and euthanasia or assisted suicide [26]. Specifically, the courts have ruled that clinicians can legally (and should, from an ethical perspective) provide patients with whatever treatments are needed to alleviate suffering (such as morphine) even if the treatments might hasten death [27]. Granting a request to withdraw life-sustaining treatments from a patient who does not want them respects the right to be left alone and to die naturally of the underlying disease, which is legally protected by the right to privacy. This has been phrased as "a right to decide how to live the rest of one's life". Finally, these rights extend to patients who lack decision-making capacity through previously expressed statements (e.g., advance directives) and/or surrogate decision-makers [22, 23, 28].

Table 11.1 Landmark legal cases confirming the right to withhold or withdraw lifesustaining therapies

In re Quinlan	1976	Supreme Court of New Jersey	Withdraw	Ventilator
Saikewicz	1977	Massachusetts Superior Judicial Court	Withhold	Chemotherapy
In the matter of Shirley Dinnerstein	1978	Massachusetts Court of Appeals	Withhold	Cardiopulmonary Resuscitation
Spring	1980	Massachusetts Superior Judicial Court	Withdraw	Hemodialysis
Barber	1983	California Court of Appeals	Withdraw	Intravenous fluids
Bouvia	1985	California Court of Appeals	Both	Feeding tube
Cruzan vs Director, Missouri Department of Health	1990	US Supreme Court	Withdraw	Feeding tube
Schiavo[a]	2005	Florida Court of Appeals[a]	Withdraw	Feeding tube

From Lampert and Hayes [2]. Reprinted with permission from Elsevier

[a]The Florida Supreme Court declined to consider case, the US Supreme Court declined to hear related case

There are healthcare providers for whom device deactivation is not consistent with their values [14, 29]. As described in the AMA Code of Medical Ethics, physicians (or other medical personnel) should not be compelled to carry out device deactivations if they view the procedure as inconsistent with their personal values. However, the clinician should not impose his or her values on or abandon the patient. Instead, the clinician and patient should work to achieve a mutually agreed-upon care plan. If such a plan cannot be achieved, then the primary clinician should involve a second clinician who is willing to co-manage the patient and provide legally permissible care and procedures including CIED deactivation [30].

Communication

A second barrier to more widespread adoption of ICD deactivation at end of life is the reluctance of physicians to discuss this option with patients. Some physicians do not feel comfortable deactivating an ICD [14, 15]. Others feel that device deactivation should be included in advanced care planning, but admit that they themselves rarely do so, finding CIED deactivation a more difficult topic than other advanced care discussions [31]. Physicians have reported lack of experience, lack of adequate rapport with patients, and time-constraints as barriers [32]. When they do discuss deactivation, it is often in the last days of a patient's life, and after they have already received unwanted shocks [1, 5].

Timely and effective communication is imperative to informed consent and to prevent unwanted shocks at the end of life. Ideally discussion of deactivation as an option should start prior to implantation and continue throughout the illness course [9]. As illness progresses, goals of care and the level of burden acceptable to a patient may change [33, 34]. Suggestions for discussion of deactivation at different points in the illness trajectory are shown in Table 11.2.

These goal-directed conversations should include discussions of quality of life, functional status, perceptions of dignity, and both current and potential future symptoms, as each of these elements can influence how patients set goals for their health care. These conversations should follow the model of "shared-decision making," in which clinicians work together with patients and families to ensure that patients understand the benefits and burdens of a particular treatment

Table 11.2 Suggested timing and phrasing for deactivation discussions [2, 9]

Timing of conversation	Points to be covered	Helpful phrases to consider
Prior to implantation	Clear discussion of the benefits and burdens of the device.	"It seems clear at this point that this device is in your best interest, but you should know at some point if you become very sick from your heart disease or another illness, the burden of the device may outweigh its benefit. While that point is hopefully a long way off, you should know that turning off your defibrillator is an option."
	Brief discussion of potential future limitations or burdensome aspects of device therapy	
	Encourage patients to have some form of advance directive	
	Inform of option to deactivate in the future	
After an episode of increased or repeated firings from an ICD	Discussion of possible alternatives, including adjusting medications, adjusting device settings, and cardiac procedures to reduce future shocks	"I know that your device caused you some recent discomfort and that you were quite distressed. I want to work with you to see if we can adjust the settings to assure that the device continues to work in the appropriate manner. If we can't get you to that point then we may want to consider turning it off altogether, but let's try some adjustments first."
Progression of cardiac disease, including repeated hospitalizations for heart failure and/or arrhythmias	Re-evaluation of benefits and burdens of device	"It appears as though your heart disease is worsening. We should really talk about your thoughts and questions about your illness at this point and see if your goals have changed at all."
	Assessment of functional status, quality of life, and symptoms	
	Referral to palliative and supportive care services	

(continued)

Table 11.2 (continued)

Timing of conversation	Points to be covered	Helpful phrases to consider
When patient/surrogate chooses a do not resuscitate order[a]	Re-evaluation of benefits and burdens of device	"Now that we've established that you would not want resuscitation in the event your heart goes into an abnormal pattern of beating, we should reconsider the role of your device. In many ways it is also a form of resuscitation. Tell me your understanding of the device and let's talk about how it fits into the larger goals for your medical care at this point."
	Exploration of patient's understanding of device and how he/she conceptualizes it with regards to external defibrillation	
	Referral to palliative care or supportive services	
Patients at end of life	Re-evaluation of benefits and burdens of device	"I think at this point we need to reconsider your [device]. Given how advanced your disease is we need to discuss whether you want to keep it active. I know this may be upsetting to talk about, but can you tell me your thoughts at this point?"
	Discussion of option of deactivation addressed with all patients, though deactivation *not* required	

Adapted from Wiegand and Kalowes [28]; with permission from Lampert and Hayes [2], and from Lampert et al. [9], both with permission from Elsevier

[a]Patients may choose to forego intubation, cardiopulmonary resuscitation, and external defibrillation while at the same time decide to keep the defibrillation function of their implantable cardioverter-defibrillator (ICD) active. A patient's choice of a "do not resuscitate" order may or may not be concomitant with a decision to withdraw device therapy, as resuscitation interventions and the ICD each carries its own benefits and burdens

and the potential outcomes that may occur as a result of its continuation or discontinuation [35]. Suggestions for communicating with patients and families regarding the role of the ICD in the context of the patient's goals of care are shown in Table 11.3.

An important role for the clinician is to provide factual and understandable information concerning the beneficial and adverse effects of continuing device therapy. Patients can then assess how the benefits and burdens of continued therapy fit with their ongoing healthcare goals. Data show that some patients with ICDs do not understand the role the device plays in their health, particularly in terms of care at the end of life [40]. An important time to discuss device deactivation is if a patient requests a "Do not resuscitate" (DNR) order, as often ICD deactivation will be concordant with current goals of care. However, there are a number of reasons why patients may wish to be "DNR", and not all patients choosing DNR will wish to have their ICD deactivated. For example, for a patient who currently has an adequate quality of life but does not want to be intubated or admitted to intensive care, leaving the ICD active may be concordant with his goals, as the negative impact of ICD shocks may be acceptable given his current quality of life. It is also important to recognize that in patients for whom ICD

Table 11.3 Steps for communicating with patients and families about goals of care relating to CIEDs [9]

	Sample phrases to use to begin conversation at each step
1. Determine what patients/ families know about their illness	"What do you understand about your health and what is occurring in terms of your illness?"
2. Determine what patients/ families know about the role the device plays in their health both now and in the future	"What do you understand the role of the [cardiac device] to be in your health?"
3. Determine what additional information patients/families want to know about their illness	"What else would you like to know about your illness or the role the [cardiac device] plays?"
4. Correcting or clarify any misunderstandings about the current illness and possible outcomes, including the role of the device	"I think you have a pretty good understanding of what is happening in terms of your health, but there are a few things I would like to clarify with you."
5. Determining the patient/ family's overall goals of care and desired outcomes	"Given what we've discussed about your health and the potential likely outcomes of your illness, tell me what you want from your health care at this point."
	NB: Sometimes patients and families may need more guidance; additional potential language might be: "At this point some patients tell me they want to live as long as possible, regardless of the outcome, whereas other patients tell me that the goal is to be comfortable for as long as possible while still being able to interact with family and friends. Do you have a sense of what you want at this point?"
6. Using the stated goals as a guide, work to tailor treatments, and in this case management of the cardiac device, to those goals.	Phrases to be used here depend on the goals set by the patient and family.
	For a patient who states that the desired goal is to live as comfortably as possible for whatever remaining time is left: "Given what you've said about assuring that you are as comfortable as possible it might make sense to deactivate the shocking function of your ICD. What do you think about that?"
	OR
	For a patient who wants all life-sustaining treatments to be continued, an appropriate response might be, "In that case, leaving the anti-arrhythmia function of the device active would be most in line with your goals. However, you should understand that if shocks occur they may cause discomfort for you and your family at the end of life. If desired, we can make a decision in the future about turning the device off. Tell me your thoughts about this."

Adapted from Refs. [36–39]; with permission from Lampert and Hayes [2], and from Lampert et al. [9], both with permission from Elsevier deactivation appears to be most concordant with their goals, individual patients may still choose to leave the device on or to defer a decision on deactivation until they receive shocks.

It is vital for both the health care provider and patient to have an accurate understanding of the expected consequences of device deactivation. In many situations it will be difficult to predict a patient's clinical course after deactivation. The timing and lethality of tachyarrhythmias is unpredictable, and elimination of shocking therapy is unlikely to result in immediate death unless the patient is experiencing incessant or increasingly frequent ventricular arrhythmias. Consultation with a clinical electrophysiologist may help clarify the clinical picture. The electrophysiologist can also ensure that the patient understands available options including changing programming to decrease likelihood of inappropriate shocks, or maintaining the anti-tachycardia pacing function (which is painless) but not the shock function.

Other members of the health care team also play a vital role in device deactivation, including nurses, social workers, and clergy. Routine psychiatric consultation is not needed for patients who are considering device deactivation but should be considered if there are concerns that a particular psychiatric disorder such as major depression or paranoid delusions may be interfering with the patient's ability to make informed decisions.

While patients and families desire conversations about end-of-life care [36, 41, 42], studies of patient perceptions of ICD deactivation have shown varying results. Several surveys have shown that only a minority of ICD patients would want their device deactivated even in hypothetical settings of advanced symptoms such as constant dyspnea or daily shocks [43, 44], and in one focus-group study, patients did not even want to discuss the possibility of device deactivation, with one patient describing it as "like an act of suicide" [40]. However, in a more recent telephone survey of 95 ICD patients over 50 years of age, 70 % would choose deactivation in one or more scenarios describing functional, cognitive, and/or medical disabilities [45]. The difference in survey findings may reflect that the more recent survey incorporated a brief informational introduction about ICD function. Patients' understanding of their ICDs has been shown in several studies to be limited [44, 45], and the fact that information provided beforehand may have impacted patients' choices underscores the importance of comprehensive communication between physicians and patients. Two studies have evaluated retrospectively what patients actually chose when asked about deactivation when they were terminally ill. In one, among eight patients with terminal cancer and ICDs, six had discussed deactivation with their physicians, and all chose to keep their ICDs active [46]. However, Lewis described results of a pro-active program to identify progression to terminal illness in ICD patients, and all of those identified (N = 20) chose deactivation [37]. It is likely that the approach used by the healthcare provider may influence these decisions, and these studies highlight the importance of effective communication of the role of the ICD in the context of the patient's overall goals of care.

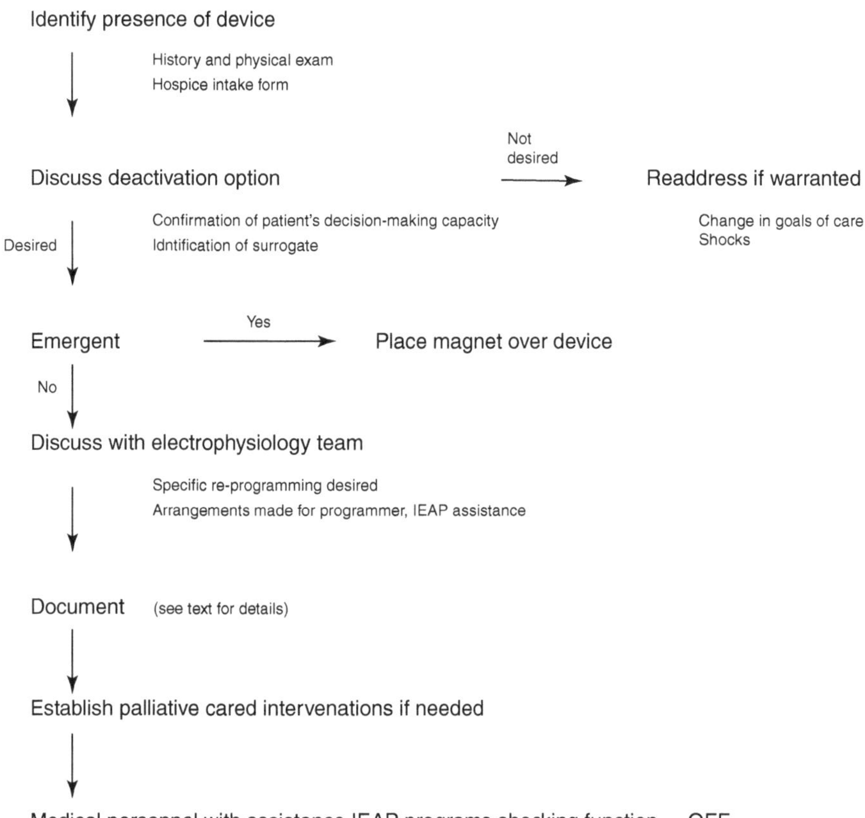

Fig. 11.1 Flow diagram for deactivation of implantable cardioverter defibrillators. *IEAP* industry-employed allied professional

Logistics

A general approach for device deactivation is shown in Fig 11.1. The most important step in giving a patient at end of life the option of ICD deactivation is recognition that a device is present. One survey of hospices revealed that only 10 % had an item regarding presence of a device on the intake form [13]. Hospices with deactivation policies in place have higher percentages of ICD patients with devices deactivated, and it has been recommended that hospices have policies in place including identification of devices on intake forms [13]. The American Heart Association has recently made similar recommendations for skilled nursing facilities [10].

When a decision for deactivation has been made, the Heart Rhythm Society recommends a series of procedures that should be consistently applied. (Adapted from Lampert et al. [9])

1. *Confirm patient's capacity to make the decision to withdraw CIED support.*
2. *Identification of the legal surrogate if the patient lacks capacity*
3. *In a long-term facility where electrophysiological expertise is not immediately available, the attending physician should contact the physician responsible for following the patient's CIED for consultation as to which therapies should be deactivated.*
4. *Documentation requirements for withdrawing a CIED:*
   Deactivation of CIED therapies requires a written order from the attending physician. This should preferably precede deactivation. In emergent situations a verbal order should be followed by written documentation within 24 h. The person responsible for ordering device deactivation may be the patient's primary care physician, cardiologist, cardiac electrophysiologist, hospitalist, or palliative care specialist. Any of these clinicians may be the most appropriate physician for such an order, though collaboration between the patient's physicians is ideal. The written documentation in the medical record should confirm:
   (a) That the patient (or legal surrogate) has requested device deactivation;
   (b) The capacity of the patient to make the decision, or identification of the appropriate surrogate;
   (c) That alternative therapies have been discussed if relevant;
   (d) That consequences of deactivation have been discussed;
   (e) The specific device therapies to be deactivated;
   (f) Notification of family if consistent with the patient's wishes.
5. *Establishing palliative care interventions and providing patient and family support.*

Device Deactivation Procedure

For patients who are well enough to travel to a clinic with programming capability an outpatient visit may be acceptable for device deactivation. However, because deactivation of some therapies may be followed by the patient's rapid demise, such as deactivation of pacing therapy in a dependent patient, the clinic setting may not always be appropriate. For patients in long-term care facilities without on-site electrophysiological expertise and who are unable to travel, deactivation should be performed by medical personnel (such as a hospice physician or nurse) with guidance from industry-employed allied professionals [47]. The attending physician should arrange for a programmer to be brought to the patient. This may require the assistance of a physician who follows CIED patients. In many cases industry-employed allied professionals (IEAPs) who represent the manufacturer of the patient's CIED will be called upon to bring a programmer to the patient's bedside. Medical personnel, ideally the attending physician, would deactivate the CIED using the programmer with the IEAP providing technical assistance. While data from a survey of Heart Rhythm Society members and IEAPs suggest that IEAPs perform deactivation 50 % of the time [14], the Heart Rhythm Society recommends [9] that the IEAP should always act under direct supervision of medical personnel except in rare emergent situations when medical personnel are not available. Communication between the electrophysiologist, IEAP, and hospice personnel is imperative to direct appropriate deactivation for patients in these facilities.

While the institution of policies designed to improve pro-active communication will reduce unwanted shocks in a dying patient, emergent situations may still occur. All ICDs can be deactivated by placing a doughnut magnet directly over the device. As devices differ in response when the magnet is removed, the magnet should be left in place until magnet function is confirmed and/or a programmer is available. All hospices should have doughnut magnets onsite and readily available, as also recommended for skilled nursing facilities [10].

Patients at home wishing device deactivation can represent a logistical challenge. For patients who are too ill to travel to a clinic or in whom deactivation would result in rapid demise, arrangements must be made for a programmer to be brought to the home by medical personnel or an IEAP. Similar to the hospice setting, deactivation should be performed by medical personnel, such as a hospice nurse, in conjunction with the IEAP. The responsible attending physician should write an order in the patient's medical record including specific therapies to be deactivated. In the rare situation in which no medical personnel can be available in a timely fashion, IEAPs may be asked to perform deactivation following appropriate communication with and documentation by the responsible attending physician. In situations where the requested deactivation is not in keeping with the manufacturer's policies, the attending physician assumes responsibility for resolving the conflict. This may involve further consultation with a physician who has expertise in CIED therapy.

Pacing and Cardiac Resynchronization Therapy

Many ICDs include the capability for resynchronization therapy (CRT), and some patients may have devices which deliver CRT alone. While ICDs prolong life without improving quality of life, CRT improves symptoms of heart failure by pacing the right and left ventricles synchronously (biventricular pacing) in patients with dyssynchronous contraction due to left bundle branch block [48, 49]. The shock function of an ICD can be disabled independently of the CRT function, and many patients may wish to leave CRT active to avoid potential worsening of heart failure symptoms.

Pacing via a standard pacemaker or an ICD is indicated to ameliorate symptoms of bradycardia [49]. Similar to CRT, patients may wish to leave pacing active to avoid potential worsening of bradycardia-related symptoms such as dyspnea or dizziness. For patients who have no underlying intrinsic rhythm (i.e., "pacemaker-dependent"), pacing also provides life-sustaining therapy. While pacemakers do not actively impair quality of life, and do not prolong the dying process, patients may determine that the benefits of the device no longer outweigh its burdens, and may request deactivation of a pacemaker or the bradycardia-pacing function of an ICD. While it has been debated whether there are moral or philosophical distinctions between pacemaker and ICD deactivation [14, 29, 50], particularly in a pacemaker-dependent patient, the principles of autonomy and the legal right to withdraw life-sustaining therapy do not depend on the device, but are based on the rights of the patient [9]. Whether dependent on the pacemaker or not, an ICD patient requesting deactivation of shocks may also wish deactivation of the pacemaker depending on his or her current goals of care and preferences, and it is imperative to discuss the outcomes of pacing deactivation and confirm the patient's wishes.

Because changing the pacing or CRT settings can result in gradual worsening of chronic symptoms or onset of new symptoms, it may be helpful to involve palliative care in the care of patients before device settings are altered. Further, palliative care clinicians can help clarify concerns and misunderstandings about device function after the patient has died. For example, families may have misperceptions that they may be shocked by touching their deceased relative, or that the device must be explanted after death (which is only true in the case of cremation). Likewise, the clinician can dispel the myth that continued pacemaker function after a patient has died is unnecessarily prolonging life, or that impulses from the device are affecting the heart in some way.

References

1. Goldstein NE, Lampert R, Bradley E, Lynn J, Krumholz HM. Management of implantable cardioverter defibrillators in end-of-life care. Ann Intern Med. 2004;141:835–8.
2. Lampert R, Hayes D. Chapter 35: ethical issues. In: Ellenbogen KA, Wilkoff BL, Kay GN, Chu-Pak Lau, editors. Clinical cardiac pacing, defibrillation and resynchronization therapy. 4th ed. Philadelphia: Elsevier/Saunders; 2011. p. 1040–9.
3. Goldberger Z, Lampert R. Implantable cardioverter-defibrillators: expanding indications and technologies. JAMA. 2006;295:809–18.
4. Hammill SC, Kremers MS, Stevenson LW, et al. Review of the registry's fourth year, incorporating lead data and pediatric ICD procedures, and use as a national performance measure. Heart Rhythm. 2010;7:1340–5.
5. Sherazi S, McNitt S, Aktas MK, et al. End-of-life care in patients with implantable cardioverter defibrillators: a MADIT-II substudy. Pacing Clin Electrophysiol. 2013;36:1273–9.
6. Ahmad M, Bloomstein L, Roelke M, Bernstein AD, Parsonnet V. Patients' attitudes toward implanted defibrillator shocks. Pacing Clin Electrophysiol. 2000;23:934–8.
7. Irvine J, Dorian P, Baker B, et al. Quality of life in the Canadian implantable defibrillator study (CIDS). Am Heart J. 2002;144:282–9.
8. Schron E, Exner D, Yao Q, et al. Quality of life in the antiarrhythmics versus implantable defibrillators. Circulation. 2002;105:589–94.
9. Lampert R, Hayes DL, Annas GJ, et al. HRS expert consensus statement on the Management of Cardiovascular Implantable Electronic Devices (CIEDs) in patients nearing end of life or requesting withdrawal of therapy. Heart Rhythm. 2010;7:1008–26.
10. Jurgens C. Heart failure management in skilled nursing facilities. Circulation. In-press.
11. Hauptman PJ, Swindle J, Hussain Z, Biener L, Burroughs TE. Physician attitudes toward end-stage heart failure: a national survey. Am J Med. 2008;121:127–35.

12. Russo JE. Original research: deactivation of ICDs at the end of life: a systematic review of clinical practices and provider and patient attitudes. Am J Nurs. 2011;111:26–35.
13. Goldstein N, Carlson M, Livote E, Kutner JS. Brief communication: management of implantable cardioverter-defibrillators in hospice: a nationwide survey. Ann Intern Med. 2010;152:296–9.
14. Mueller PS, Jenkins SM, Bramstedt KA, Hayes DL. Deactivating implanted cardiac devices in terminally ill patients: practices and attitudes. Pacing Clin Electrophysiol. 2008;31: 560–8.
15. Sherazi S, Daubert JP, Block RC, et al. Physicians' preferences and attitudes about end-of-life care in patients with an implantable cardioverter-defibrillator. Mayo Clin Proc. 2008;83: 1139–41.
16. Beauchamp TL. Principles of biomedical ethics. 6th ed. New York: Oxford University Press; 2009.
17. Snyder L, Leffler C. Ethics manual: fifth edition. Ann Intern Med. 2005;142:560–82.
18. AMA Council on Ethical and Judicial Affairs. AMA 2008–2009. Code of medical ethics: current opinions and annotations. 2008–2009 edition. Chicago: AMA Press; 2010.
19. Annas GJ. The rights of patients: the authoritative ACLU guide to the rights of patients. 3rd ed. New York: New York University Press; 2004.
20. Pellegrino ED. Decisions to withdraw life-sustaining treatment: a moral algorithm. JAMA. 2000;283:1065–7.
21. Rhymes JA, McCullough LB, Luchi RJ, Teasdale TA, Wilson N. Withdrawing very low-burden interventions in chronically ill patients. JAMA. 2000;283:1061–3.
22. Quill TE, Barold SS, Sussman BL. Discontinuing an implantable cardioverter defibrillator as a life-sustaining treatment. Am J Cardiol. 1994;74:205–7.
23. Annas GJ. "Culture of life" politics at the bedside – the case of Terri Schiavo. N Engl J Med. 2005;352:1710–5.
24. Burt RA. Death is that man taking names. Berkeley: University of California Press; 2002.
25. Schneider C. The practice of autonomy: patients, doctors, and medical decisions. New York: Oxford University Press; 1998.
26. Vacco v Quill. 521 U.S. 793, 95–1858. Supreme Court of the United States; 1997.
27. Washington v. Glucksberg. 521 U.S. 702, 96–110. Supreme Court of the United States; 1997.
28. Wiegand DL, Kalowes PG. Withdrawal of cardiac medications and devices. AACN Adv Crit Care. 2007;18:415–25.
29. Kay GN, Bittner GT. Should implantable cardioverter-defibrillators and permanent pacemakers in patients with terminal illness be deactivated? Deactivating implantable cardioverter-defibrillators and permanent pacemakers in patients with terminal illness. An ethical distinction. Circ Arrhythm Electrophysiol. 2009;2:336–9.
30. AMA Council on Ethical and Judicial Affairs. Physician objection to treatment and individual patient discrimination: CEJA report 6-A-07. AMA council on ethical and judicial affairs. Chicago: AMA Press; 2007.
31. Goldstein NE, Mehta D, Teitelbaum E, Bradley EH, Morrison RS. "It's like crossing a bridge" complexities preventing physicians from discussing deactivation of implantable defibrillators at the end of life. J Gen Intern Med. 2008;1:2–6.
32. Kelley AS, Reid MC, Miller DH, Fins JJ, Lachs MS. Implantable cardioverter-defibrillator deactivation at the end of life: a physician survey. Am Heart J. 2009;157:702–8.e1.
33. Fried TR, Bradley EH, Towle VR, Allore H. Understanding the treatment preferences of seriously ill patients. N Engl J Med. 2002;346:1061–6.
34. Fried TR, Byers AL, Gallo WT, et al. Prospective study of health status preferences and changes in preferences over time in older adults. Arch Intern Med. 2006;166:890–5.
35. Goldstein NE, Back AL, Morrison RS. Titrating guidance: a model to guide physicians in assisting patients and family members who are facing complex decisions. Arch Intern Med. 2008;168:1733–9.
36. Fried TR, O'Leary JR. Using the experiences of bereaved caregivers to inform patient- and caregiver-centered advance care planning. J Gen Intern Med. 2008;23:1602–7.

37. Lewis WR, Luebke DL, Johnson NJ, Harrington MD, Costantini O, Aulisio MP. Withdrawing implantable defibrillator shock therapy in terminally ill patients. Am J Med. 2006;119:892–6.
38. Lynn J, Goldstein NE. Advance care planning for fatal chronic illness: avoiding commonplace errors and unwarranted suffering. Ann Intern Med. 2003;138:812–8.
39. Morrison RS, Meier DE. Clinical practice. Palliative care. N Engl J Med. 2004;350:2582–90.
40. Goldstein NE, Mehta D, Siddiqui S, et al. "That's like an act of suicide" patients' attitudes toward deactivation of implantable defibrillators. J Gen Intern Med. 2008;1:7–12.
41. Singer PA, Martin DK, Kelner M. Quality end-of-life care: patients' perspectives. JAMA. 1999;281:163–8.
42. Nicolasora N, Pannala R, Mountantonakis S, et al. If asked, hospitalized patients will choose whether to receive life-sustaining therapies. J Hosp Med. 2006;1:161–7.
43. Raphael CE, Koa-Wing M, Stain N, Wright IAN, Francis DP, Kanagaratnam P. Implantable cardioverter-defibrillator recipient attitudes towards device deactivation: How much do patients want to know? Pacing Clin Electrophysiol. 2011;34:1628–33.
44. Stewart GC, Weintraub JR, Pratibhu PP, et al. Patient expectations from implantable defibrillators to prevent death in heart failure. J Cardiac Fail. 2010;16:106–13.
45. Dodson JA, Fried TR, Van Ness PH, Goldstein N, Lampert R. Patient preferences for deactivation of implantable cardioverter defibrillators. JAMA Intern Med. 2013;173:377–9.
46. Kobza R, Erne P. End of life decisions in patients with malignant tumors. Pacing Clin Electrophysiol. 2007;30:845–9.
47. Lindsay BD, Estes 3rd NA, Maloney JD, Reynolds DW, Heart RS. Heart rhythm society policy statement update: recommendations on the role of Industry Employed Allied Professionals (IEAPs). Heart Rhythm. 2008;5.
48. Bristow MR, Saxon LA, Boehmer J, et al. Cardiac-resynchronization therapy with or without an implantable defibrillator in advanced chronic heart failure. N Engl J Med. 2004;350:2140–50.
49. Epstein AE, DiMarco JP, Ellenbogen KA, et al. ACC/AHA/HRS 2008 guidelines for device-based therapy of cardiac rhythm abnormalities: a report of the American College of Cardiology/American Heart Association Task Force on Practice Guidelines (writing committee to revise the ACC/AHA/NASPE 2002 guideline update for implantation of cardiac pacemakers and antiarrhythmia devices): developed in collaboration with the American Association for Thoracic Surgery and Society of Thoracic Surgeons. Circulation. 2008;117:e350–408.
50. Zellner RA, Aulisio MP, Lewis WR. Deactivating permanent pacemakers in patients with terminal illness; patient autonomy is paramount. Circ Arrhythm Electrophysiol. 2009;5(3): 340–4.

Chapter 12
Models of End-of-Life Care in the Home Environment

Susan Enguidanos and Richard D. Brumley

Abstract People with advanced heart failure spend most of their time living in the community, therefore could benefit by improved access to palliative care in the home. Hospice and home-based palliative care are the primary mechanisms for provision of such care for community-dwelling patients with heart failure. While several barriers to hospice care exist for heart failure patients, significant evidence for improved patient outcomes for home-based palliative care exists. Improved patient satisfaction as well as physical and psychological symptom control has been demonstrated along with considerable reductions in costs of medical care for patients with advance heart disease enrolled in home-based palliative acre programs. Despite this evidence for effectiveness among patients, caregivers of heart failure patients may continue to experience high levels of burden.

Keywords Advanced heart failure • Home-based palliative care • Hospice • Home care

Key Points

- Hospice care can provide improved access to pain relief, however, few patients with heart failure receive hospice.
- Home-based palliative care can improve pain and symptoms, including anxiety and depression, for patients with advanced heart failure.
- In addition to improved patient outcomes, home-based palliative care has demonstrated ability to reduce medical care costs.
- Despite provision of support and resources, caregivers of patients with heart failure receiving home-based palliative care may continue to experience high levels of burden and sleep disturbances.

S. Enguidanos, MPH, PhD (✉)
Leonard Davis School of Gerontology, University of Southern California,
3715 McClintock Ave, Gero 208B, Los Angeles, CA 90089-0191, USA
e-mail: Enguidan@usc.edu

R.D. Brumley, MD
Hospice and Palliative Medicine, Laguna Niguel, CA, USA

S.J. Goodlin, M.W. Rich (eds.), *End-of-Life Care in Cardiovascular Disease*,
DOI 10.1007/978-1-4471-6521-7_12

The Need for Palliative Care in the Home

At any given point, approximately 95 % of older persons reside in the community rather than in institutions [1, 2]. Further, the majority of older adults prefer to remain in the community for as long as possible [3–5]. Early studies on end of life estimated that terminally ill patients spend the vast majority of their last year of life in the community rather than in the hospital or other institutions [6, 7]. A more recent study in England found that the majority of patients spend the last year of life at home under the care of their own family physician [8]. Thus, provision of palliative care within the predominant environment is a practical response to improving the availability of palliative care and ensuring continuity of care.

While hospital-based palliative care provides timely intervention and services in the acute care setting, it leaves minimal opportunities for follow-up unless the patient is eligible and has enrolled in hospice. In addition, many seriously ill patients face barriers to accessing primary care or outpatient palliative clinic visits, including lack of transportation and mobility challenges. For many terminal illnesses, the end-of-life trajectory is riddled with complex symptom management issues, problems and difficulty in ambulation, and fatigue and weakness [9]. Physical decline can serve as a barrier to primary care services for seriously ill patients who either do not qualify for hospice or do not want those services [10, 11]. This point is eloquently illustrated in a passage by Bookvar (p. 2246) [9]:

> Dr. T: 'She's had a very long history of heart disease … I would always marvel that she could manage to [come to clinic to see me] over the last year … She's very frail but able to get around at home and do the things she wants to do.
>
> In July 2005, noticing Mrs. K's declining status and progressive weight loss, Dr. T broached the possibility of hospice care. She and her son were unsettled by the hospice worker's discussion of end-of-life care and hospice services, so Mrs. K declined.

Moreover, provision of care within the home setting allows for a full understanding of both barriers and facilitators in the ability of the patient to receive adequate care, factors that would not be apparent without first hand observation of the home resources [12]. These factors include

- home safety issues
- caregiver functioning
- Patient functioning in the actual home environment
- Other equipment and resource needs.

HBPC also supports training of the patient and caregiver within the context of their everyday environment. While instruction can be provided in the hospital or physician office, these training opportunities generally present at times of acute stress for the patient. Furthermore, the information provided must later be recalled and translated to the home environment. Instructions provided in the home can be individualized, tailored to the home environment and reinforced at subsequent visits. This ensures that the intervention or treatment is understood and facilitates the family and caregiver's ability to implement the care plan.

Finally, the home is a place of comfort. Being at home allows people to remain close to family, in familiar surroundings and living as normally as possible. This location also enables the patient and family to be in control of care and central to care choices. Stewart [13] makes a convincing case for provision of care for HF patients in the home, listing the myriad physical and psychological challenges encountered as well as a complicated set of health care needs and behaviors that can be better identified and managed in the home environment [13].

One of the specific aims of palliative care is to provide improved quality of care at end of life, with increasing emphasis on identifying and respecting patient wishes and goals [14]. Most older adults prefer to reside in the community and die at home [4, 15, 16]. A nationwide study found that 82 % of older adults would prefer to remain in their homes even in the event they should require help caring for themselves [17]. Although some studies have documented the variability in patient preference for site of death [16, 18, 19], population polls and studies overwhelmingly have found that most people also prefer home as the site of death [15, 16, 20]. Reversal of preference for home death has been associated with fears of vulnerability, unmanaged symptoms, and concern about family burden [21].

Family members of individuals dying at home and under the care of hospice report greater emotional support and quality of care [22]. Similarly, patients dying in the hospital have reported poorer quality of life and greater physical and emotional distress as compared to those dying at home under the care of hospice [23]. Thus, the availability of palliative care programs within the home environment is important to ensure continuity of quality care at end of life.

Hospice for HF Patients

Hospice care has been a Medicare benefit since 1985, initially primarily serving patients with cancer. Admission requirements for hospice care include having a terminal medical condition with an estimated 6 month prognosis. For patients with HF, hospice guidelines include that the patient is optimally treated, has discomfort with activity (New York Heart Association Class IV), is symptomatic at rest (e.g. angina), and has an ejection less than or equal to 20 % if available. Additionally, patients must agree to forgo curative care measures. Since its inception, enrollment in hospice care has increased dramatically and expanded to include a wider range of conditions. Although heart disease is the leading cause of death in the US, accounting for 23.7 % of all deaths in 2011 [24], only 11.2 % of hospice recipients have heart disease [25]. However, among those with HF that are admitted to hospice care, they have a higher average length of stay as compared to cancer patients and are more likely to be discharged alive, perhaps due to the unpredictable disease trajectory [26]. About two-thirds (66 %) of hospice care is provided in the patient's home [25]. Moreover, hospice care has been found to be positively associated with greater receipt of pain medications among heart failure and cancer patients [27].

Further limiting hospice access are variations in individual hospice agency enrollment criteria that are not aligned with the Medicare benefit. In fact, one study found that 78 % of hospice agencies in a national survey had at least one policy that decreased access to hospice service [28]. These policies include prohibiting:

- Any chemotherapy (including palliative chemotherapy)
- Parenteral nutrition or tube feeding
- Transfusions
- Intrathecal catheter
- Palliative radiation
- Hospice care for those without a caregiver at home

Home-Based Palliative Care

Despite national guidelines and recommendations for palliative care for heart failure patients, many individuals with heart failure do not receive hospice care. The unpredictable trajectory of HF and the low enrollment in hospice among patient with heart disease demonstrates the need for other models of palliative care to better support the myriad of palliative care needs experience by patients with heart failure.

One such model is home-based palliative care (HBPC). HBPC aims to provide coordinated, comprehensive palliative care upstream in the disease trajectory (see Fig. 12.1). Most HBPC models provide care concurrent with usual care, including curative and aggressive care interventions [29–31]. Individuals receive HBPC based on symptom severity, rather than life expectancy as required by the hospice benefit [32]. For example, the HBPC program developed and implemented within Kaiser Permanente in southern California targeted heart failure patients with a palliative performance scale of 50 % or less and a recent history of emergency room and

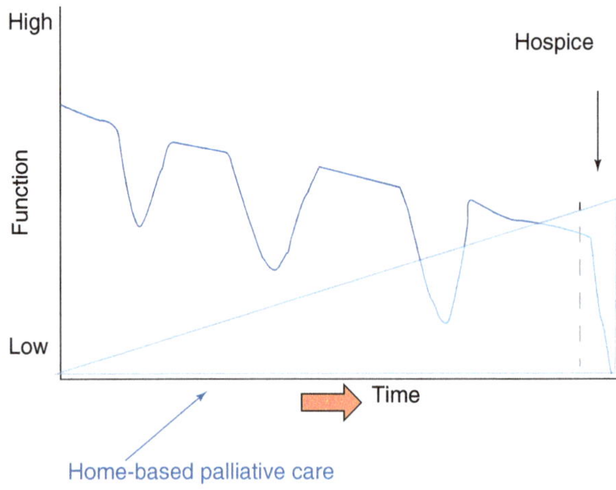

Fig. 12.1 Heart failure trajectory and palliative care models

hospital visits, a marker of declining health and rising health care needs. Additionally, the primary care physician was asked if they would be surprised if the patient died in the next year [33, 34].

While HBPC models vary, much of the care and HBPC team composition mirror that of hospice. The interdisciplinary care team consists of a physician, nurse, social worker and allied professionals from other disciplines, including clergy, physical therapy, music therapy, and home health aides. Education provided by the team on disease progression and symptom management conveys critical information that enables patients and families to make informed care decisions and allows them to plan and prepare for symptom exacerbation and end-of-life care choices. Also key to this model is the focus on symptom management with support from a 24 h, 7-day-a-week call center. Through this readily available support system, seriously ill patients and their families can receive professional assistance whenever the need arises.

All HBPC services are provided in the home. The palliative care team begins by identifying the medical, psychosocial, and spiritual needs of the patient and family as part of a comprehensive interdisciplinary assessment. Working with the patient and family, the team develops an individualized care plan that outlines treatment, palliative, psychosocial, spiritual, and other goals in the care of the patient. The subsequent rate and frequency of visits is based on the care plan and the strides made in achieving the goals elicited. For example, the care plan may specify several nursing visits per week for a period of 2 weeks to stabilize pain and symptoms. Once this is accomplished, nursing visits may decrease to one or two a month or as necessary to maintain stability and compliance to care plan. Similarly, social work visits may begin with two per week to address family conflict issues around end-of-life care decisions, decreasing to one per month after resolution.

Although HBPC and hospice programs have similar features, the HBPC model varies from hospice care in several ways. First, care is provided further upstream in the course of treatment, with focus on the last year of life rather than the last 6 months. This allows earlier enrollment and increased benefit from palliative care. As a result of advance care planning, education, training in intervention techniques, and 24-h access to medical advice and care, the likelihood of crisis is reduced. Second, patients enrolled in Home-based Palliative Care are not required to make potentially difficult decisions to forgo curative efforts, as is required by the hospice program. Similar to hospice care, patients can maintain their primary and specialty physicians, however, under HBPC, patients also may continue to pursue aggressive measures. Health care is enhanced through coordination of care and care plans by the palliative care team with the primary care specialty physicians. Additionally, enrollment in the HBPC program is not meant as a substitute for hospice care. As the disease advances and symptoms worsen, the palliative care team provides referrals to hospice services in response to patients' changing needs. Additionally, some patients may meet all criteria for hospice but simply do not want to be on a hospice program. The HBPC program can serve as a substitute for hospice for these patients, many of whom will elect hospice at a later time (see Table 12.1).

Table 12.1 Comparison of hospice versus home-based palliative care

Hospice care	Home-based palliative care
Provided in last 6 months of life	Provided in last year or two of life
Forego curative care; focus of care is palliative	May pursue curative care concurrent with palliative care
Medicare and most insurances pay for all hospice care	Select palliative care services may be covered by Medicare and other insurances

While recognizing that most individuals prefer to die at home, the HBPC program is also designed to take into account that not everyone has the same preferences for site of death. Thus, identifying and incorporating these preferences into the care plan is essential to the model. For cultural and personal reasons, some people are uncomfortable with having a death in the house and request transfer to a hospital or nursing facility when death is imminent. Since the primary objective of the HBPC program is to provide care consistent with patient and family wishes, care plans are established with the intent of following patient preferences.

As with preferred site of death, other life sustaining interventions are documented according to the desires of the patients. The goal of the palliative care intervention isn't to "convert" the patient and family toward less aggressive care, even when aggressive care may appear futile to team members. Rather, the palliative care team provides comprehensive information on likely outcomes of life sustaining interventions, such as feeding tubes and resuscitation, including data from clinical trials and meta-analysis. In this way, patients and caregivers have access to all available information required to make an informed decision. As with hospice care, HBPC patients may elect to receive CPR, however under HBPC patients are not limited in the life sustaining interventions they seek. The goal of the HBPC team is to "honor patient preferences" once the patient and family have been educated about the likely impact of alternate care choices. See Text Box below for a case example of HBPC for a heart failure patient.

Home-Based Palliative Care Case Study

Robert Lewis, a 79 year-old male, was recently hospitalized for the third time in 6 months for an acute exacerbation of his heart failure (HF). He has been followed by his primary care physician for over 20 years but has found it difficult to keep outpatient appointments in the last year. Robert feels his HF has been under good control on Furosemide 20 mg daily. He also has diabetes mellitus and rarely checks his blood sugars but feels his diabetes is under good control as he is "never thirsty." He had gained 8 lb in the 2 months between his previous hospital discharge and the recent admission.

Robert helps care for his 78 year-old wife, Marie, who has early signs of Alzheimer's dementia. The couple lives in their home of 48 years. Their only relative is their daughter who lives in Michigan and hasn't visited for the last 3 years. With his wife's decline over the past few months, Robert has found it increasingly difficult to manage daily living activities. A neighbor volunteers to purchase prepared meals for the couple and looks in on them a few times a week. Robert wonders how he and Marie will manage as they get older and "don't want to be a burden to their daughter."

Care Opportunities

Hospice vs. Palliative Care. During his most recent hospitalization, the hospitalist suggested that Robert could qualify for hospice, but Robert felt he was not ready for hospice. A home-based palliative care (HBPC) program was offered to him with periodic home visits by a physician, nurse and social worker, which Robert accepted because he felt this would be beneficial.

Medical Management. The HBPC team instructed Robert on how to manage his heart failure. Although he had been told to check his weight daily, he never did this; his scale was old and he was unable to find it. The HBPC team encouraged him to buy a new scale and record his weight daily. If his weight increases more than 2 lb in 1 day or 3 lb in a week, he was instructed to notify the HBPC team. Since he still had 1+ edema in his ankles and a few scattered rales in the base of his lungs, the HBPC team determined his HF to be mild. His Furosemide was increased to 40 mg daily. His target or "dry" weight was determined to be 145 lb, with his current weight at 148 lb. Thus a goal was set to reduce his weight another 3 lb or until he has no more edema or rales. Although is current blood sugars were found to be satisfactory, Robert was instructed to check and record his random and fasting blood sugars a few times a week for review by the HBPC team.

Code Status. During his last hospitalization, the inpatient palliative care team met with Robert to discuss his concerns, goals, and aspirations for future care. He elected a "no code" and completed a POLST form indicating his desire for comfort care and a natural death.

Dietary Issues. Robert relied on his neighbor to purchase prepared meals for him and Marie. The HBPC nurse informed him that prepared meals are often high in sodium, which aggravates his HF. He met with a nutritionist who offered basic dietary counseling and the HBPC social worker provided a referral for home delivered meals, a healthier food option.

Social Issues. Robert's concern about long-term care for himself and his wife, especially as her Alzheimer's progresses, was addressed by the HBPC social worker who suggested they discuss future care needs with their daughter. His daughter agreed to visit in several weeks at which time the social worker will hold a family conference. Robert did state that he and Marie prefer to remain in their home, and that he would consider hiring someone to assist, at least a few hours a day to start.

Home Safety. The HBPC nurse discussed concerns she noted in the patient's home including extension cords across walkways, throw rugs with uneven borders, and the need for grab bars in the bathroom.

Current Status

Robert has been receiving HBPC for 6 months. His health is stable and he has not been hospitalized. Although his wife's health has continued to decline, they are managing well with the aid of a hired caregiver 5 days a week, paid for in part by Robert's daughter.

Evidence for Improved Outcomes Among HF Patients

There is growing evidence supporting improved outcomes for patients with heart failure receiving HBPC. A recent study of home-based primary and palliative care for seriously ill homebound patients with heart failure and other conditions found that interdisciplinary palliative care visits along with home visits by specialists resulted in lower rates of pain, anxiety and depression, tiredness, and lack of appetite following enrollment [35]. Patient satisfaction with care has also been demonstrated in studies of HBPC [30], with those enrolled in HBPC reporting greater satisfaction as compared with individuals receiving usual care.

A nurse practitioner model of HBPC for HF and COPD patients in their last 2 years of life or less found patients receiving palliative care were better prepared to manage their health condition (and for, or at) end of life [36]. Intervention patients reported higher levels of preparedness for emergencies, having more information to handle them and knowing who to contact as compared with patients receiving usual care. However, HF patients receiving the nurse practitioner intervention reported greater distress than controls, perhaps an indicator of the need for a more intensive home intervention [36].

Studies have also demonstrated that HBPC can improve rates of home deaths for heart failure patients, an important outcome given that most adults prefer to die at home [15, 16, 20]. One study found that 87 % of HBPC patients died at home as compared to only 47 % of heart failure patients receiving usual care [37]. In multivariate models, those receiving HBPC were eight times more likely to die at home than usual care patients [37]. Meta-analysis of HBPC studies found that receipt of HBPC more than doubles the odds of dying at home [20].

Evidence for Reduction in Health Service Use

Evidence for reduction in medical service use among patients receiving HBPC is considerable. Interdisciplinary HBPC teams have documented reduction in emergency room visits hospital stays, with about 40 % fewer patients using the emergency room and hospitalized than those receiving standard medical care [30, 38]. These reductions in costly medical service use have resulted in overall reductions in health service ranging from 33 to 45 %. In examining HF specific outcomes, reductions in overall health care costs are even higher, with analysis finding costs of care for HBPC patients with HF 52 % lower than HF patients receiving usual care [37]. A HBPC consultation model using a nurse practitioner to provide home visits and collaborate with primary care also found fewer hospitalizations, hospital days, and lower overall costs of medical care in the 18 months following enrollment in care [39].

Receipt of HBPC has been found to reduce 30-day hospital readmissions among seriously ill patients, including those with heart failure [39, 40]. Among hospitalized patients receiving an inpatient palliative care consultation and subsequently discharged to home, those enrolled in home-based palliative care (8 % readmitted) or hospice (5 % readmitted) were significantly less likely to be readmitted than those discharged to home with no care (26 % readmitted) or a nursing facility (25 % readmitted) [40].

Evidence for Caregiver Outcomes

Despite considerable evidence attesting to the positive outcomes for patients, similar experiences for caregivers have not been found. In fact, several studies examining the caregiver experience among heart failure patients receiving HBPC point to high burden for caregivers. Although caregivers report HBPC provides added support and shared responsibility [41], stress and burden persist. Caregivers are impacted by the unpredictability of HF, the physical and psychological challenges of being on constant alert and ready to intervene [41]. Sleep disturbances among caregivers of heart failure patients have been well-documented as well [42, 43].

Further, dyspnea or shortness of breath is a common symptom experienced by patients with advanced heart failure [44] and a difficult symptom for caregivers to manage [45]. Dyspnea is associated with greater levels of sleep disturbances and fewer positive experiences as a caregiver [42]. Although palliative care interventions can not completely eliminate symptoms of dyspnea, a HBPC team can provide support for caregivers in handling dyspnea. Caregivers commonly experience a sense of inadequacy and helplessness in dealing with episodes of dyspnea, however, support and shared responsibility can mitigate these feelings and provide strength for the caregiver [45].

References

1. AOA. A profile of older Americans: 2011. AOA; Washington, 2011. http://www.aoa.gov/AoAroot/Aging_Statistics/Profile/2011/docs/2011profile.pdf. http://www.acl.gov/About_ACL/Contact_Us/Index.aspx.
2. AARP. Independent living: progress in the housing of older persons: AARP public policy institute. AARP; Washington, 1999. http://www.aarp.org/home-garden/housing/info-1999/aresearch-import-150-D16953.html.
3. Farber N, Shinkle D, Lynott J, Fox-Grage W, Harrell R. Aging in place: a state survey of livability policies and practices, vol. 190. AARP; Washington, 2011. http://assets.aarp.org/rgcenter/ppi/liv-com/ib190.pdf.
4. Eckert JK, Morgan LA, Swamy N. Preferences for receipt of care among community-dwelling adults. J Aging Soc Policy. 2004;16(2):49–65. doi:10.1300/J031v16n02_04. http://dx.doi.org.libproxy.usc.edu/10.1300/J031v16n02_04.
5. McAuley WJ, Blieszner R. Selection of long-term care arrangements by older community residents. Gerontologist. 1985;25(2):188–93.
6. Long SH, Gibbs JO, Crozier JP, Cooper Jr DI, Newman Jr JF, Larsen AM. Medical expenditures of terminal cancer patients during the last year of life. Inquiry. 1984;21(4):315–27. http://www.jstor.org.libproxy.usc.edu/stable/29771662.
7. McCall N. Utilization and costs of medicare services by beneficiaries in their last year of life. Med Care. 1984;22(4):329–42. http://www.jstor.org.libproxy.usc.edu/stable/3765035.
8. Higginson IJ, Astin P, Dolan S. Where do cancer patients die? Ten-year trends in the place of death of cancer patients in England. Palliat Med. 1998;12:353–63.
9. Boockvar KS, Meier D. Palliative care for frail older adults: there are things I can't do anymore that I wish I could. JAMA. 2006;296(18):2245.
10. Fitzpatrick AL, Powe NR, Cooper LS, Ives DG, Robbins JA. Barriers to health care access among the elderly and who perceives them. Am J Public Health. 2004;94(10):1788–94.
11. Fried T, Wachtel T, Tinetti M. When the patient cannot come to the doctor: a medical housecalls program. J Am Geriatr Soc. 1998;46(2):226–31.
12. Lawson R. Home and hospital; hospice and palliative care: how the environment impacts the social work role. J Soc Work End Life Palliat Care. 2007;3(2):3–17.
13. Stewart S. What is the optimal place for heart failure treatment and care: home or hospital? Curr Heart Fail Rep. 2013;10:227–31.
14. Meier D. Palliative care in hospitals. J Hosp Med. 2006;1(1):21–8.
15. Hays JC, Galanos AN, Palmer TA, McQuoid DR, Flint EP. Preference for place of death in a continuing care retirement community. Gerontologist. 2001;41(1):123–8.
16. Tang ST. Determinants of hospice and home care use among terminally ill cancer patients. Nurs Res. 2003;52(4):217–25.
17. AARP. Fixing to stay: a national survey on housing and home modification issues research report. AARP; Washington, 2000. http://assets.aarp.org/rgcenter/il/home_mod.pdf.
18. Evans WG, Cutson TM, Steinhauser KE, Tulsky JA. Is there no place like home? caregivers recall reasons for and experience upon transfer from home hospice to inpatient facilities. Palliat Med. 2006;9(1):100–10.
19. Fried TR, van Doorn C, O'Leary JR, Tinetti ME, Drickamer MA. Older persons' preferences for site of terminal care. Ann Intern Med. 1999;131(2):109–12.
20. Gomes B, Calanzani N, Curiale V, McCrone P, Higginson IJ. Effectiveness and cost-effectiveness of home palliative care services for adults with advanced illness and their caregivers. Cochrane Database Syst Rev. 2013;6.
21. Munday D, Petrova M, Dale J. Exploring preferences for place of death with terminally ill patients: qualitative study of experiences of general practitioners and community nurses in England. Br Med J. 2009;339:b2391.
22. Teno JM, Clarridge BR, Casey V, et al. Family perspectives on end-of-life care at the last place of care. JAMA. 2004;291(1):88–93.

23. Wright AA, Keating NL, Balboni TA, Matulonis UA, Block SD, Prigerson HG. Place of death: correlations with quality of life of patients with cancer and predictors of bereaved caregivers' mental health. J Clin Oncol. 2010;28(29):4457–64.
24. Hoyert DL, Xu J. Deaths: preliminary data for 2011. National vital statistics report, 6th ed. vol. 61. National Center for Health Statistics: Hyattsville; 2012.
25. NHPCO. NHPCO's facts and figures: hospice care in America. NHPCO; Alexandria, 2013. http://www.nhpco.org/sites/default/files/public/Statistics_Research/2013_Facts_Figures.pdf.
26. Unroe KT, Greiner MA, Hernandez AF, et al. Resource use in the last 6 months of life among medicare beneficiaries with heart failure, 2000–2007. Arch Intern Med. 2011;171(3):196–203.
27. Miller SC, Mor V, Wu N, Gozalo P, Lapane K. Does receipt of hospice care in nursing homes improve the management of pain at the end of life? J Am Geriatr Soc. 2002;50(3):507–15. doi:10.1046/j.1532-5415.2002.50118.x.
28. Carlson MDA, Barry CL, Cherlin EJ, McCorkle R, Bradley EH. Hospices' enrollment policies may contribute to underuse of hospice care in the united states. Health Aff. 2012;31(12):2690–8. http://search.proquest.com.libproxy.usc.edu/docview/1242469445?accountid=14749.
29. Brännström M, Boman K. A new model for integrated heart failure and palliative advanced homecare – rationale and design of a prospective randomized study. Eur J Cardiovasc Nurs. 2012. doi:10.1177/1474515112445430.
30. Brumley R, Enguidanos S, Jamison P, et al. Increased satisfaction with care and lower costs: results of a randomized trial of in-home palliative care. J Am Geriatr Soc. 2007;55(7):993–1000.
31. Klinger CA, Howell D, Marshall D, Zakus D, Brazil K, Deber RB. Resource utilization and cost analyses of home-based palliative care service provision: the Niagara West End-of-Life Shared-Care Project. Palliat Med. 2013;27(2):115–22. doi:10.1177/0269216311433475.
32. Adler ED, Goldfinger JZ, Kalman J, Park ME, Meier DE. Contemporary reviews in cardiovascular medicine: palliative care in the treatment of advanced heart failure. Circulation. 2009;120:2597–606.
33. Weiche R, Mundy B, Skokan L, Shivers C, Simmons A, Gordon SG. Identifying patients nearing the end of life from congestive heart failure or chronic obstructive pulmonary disease. J Gen Intern Med. 2000;15:9–10.
34. Lynn J, Schuster JL, Kabcenell A. Improving care for the end of life: a sourcebook for health care managers and clinicians. New York: Oxford Press; 2000.
35. Ornstein K, Wajnberg A, Kaye-Kauderer H, et al. Reduction in symptoms for homebound patients receiving home-based primary and palliative care. J Palliat Med. 2013;16(9): 1048–54.
36. Aiken LS, Butner J, Lockhart CA, Volk-Craft BE, Hamilton G, Williams FG. Outcome evaluation of a randomized trial of the PhoenixCare intervention: program of case management and coordinated care for the seriously chronically ill. Palliat Med. 2006;9(1):111–26.
37. Enguidanos SM, Cherin D, Brumley R. Home-based palliative care study. J Soc Work End Life Palliat Care. 2005;1(3):37–56. doi:10.1300/J457v01n03_04. http://dx.doi.org.libproxy.usc.edu/10.1300/J457v01n03_04.
38. Brumley R, Enguidanos SM, Cherin D. Effectiveness of a home-based palliative care program for end-of-life. J Palliat Med. 2003;6(5):715–24.
39. Lukas L, Foltz C, Paxton H. Hospital outcomes for a home-based palliative medicine consulting service. J Palliat Med. 2013;16(2):179–84.
40. Enguidanos S, Vesper E, Lorenz K. 30-day readmissions among seriously ill older adults. J Palliat Med. 2012;15(12):1356–61.
41. Brännström M, Ekman I, Boman K, Strandberg G. Being a close relative of a person with severe, chronic heart failure in palliative advanced home care ? A comfort but also a strain. Scand J Caring Sci. 2007;21(3):338–44. doi:10.1111/j.1471-6712.2007.00485.x.
42. Malik FA, Gysels M, Higginson IJ. Living with breathlessness: a survey of caregivers of breathless patients with lung cancer or heart failure. Palliat Med. 2013;27(7):647–56. doi:10.1177/0269216313488812.

43. Carlsson ME. Sleep disturbance in relatives of palliative patients cared for at home. Palliat Support Care. 2012;10(03):165–70. doi:10.1017/S1478951511000836. http://dx.doi.org/10.1017/S1478951511000836.
44. Levenson JW, McCarthy EP, Lynn J. The last six months of life for patients with congestive heart failure. J Am Geriatr Soc. 2000;48(5):S101–9. http://search.ebscohost.com/login.aspx?direct=true&db=ssf&AN=510090367&site=ehost-live.
45. Gysels MH, Higginson IJ. Caring for a person in advanced illness and suffering from breathlessness at home: threats and resources. Palliat Support Care. 2009;7(02):153–62. doi:10.1017/S1478951509000200. http://dx.doi.org/10.1017/S1478951509000200.

Chapter 13
Care for Patients Dying with a Left Ventricular Assist Device

Justin M. Vader and Susan M. Joseph

Abstract Left ventricular assist devices (LVADs) are increasingly being used to support patients with advanced heart failure, both as a bridge to heart transplantation and as a final or 'destination' therapy. Compared with continued medical therapy alone, these devices prolong survival and improve overall patient functional status and quality of life – as such LVADs are not a strictly palliative therapy. However, life with LVAD support is frequently accompanied by device-related complications such as bleeding, ventricular arrhythmias, infection, stroke, renal failure, and right ventricular failure. Management of the LVAD-supported patient who has experienced device-related complications requires both a focus on the underlying pathophysiology and the patient's overall goals of care. Palliative or supportive care approaches serve an important complementary role in the management of these patients and may also be used to inform decision-making prior to device implantation. We present a review of the complications of LVAD support and their treatment, describe a role for palliative care in the management of these patients, and consider future directions for research to improve the care of LVAD-supported patients.

Keywords Advanced heart failure • End-of-life care • Left ventricular assist device • Mechanical • Circulatory support • Palliative care

---

J.M. Vader, MD • S.M. Joseph, MD (✉)
Cardiovascular Division, Department of Internal Medicine, Washington University in St Louis, 660 S. Euclid Ave., 8086, St. Louis, MO 63110, USA
e-mail: sjoseph@dom.wustl.edu

S.J. Goodlin, M.W. Rich (eds.), *End-of-Life Care in Cardiovascular Disease*,
DOI 10.1007/978-1-4471-6521-7_13

Key Points

- Left ventricular assist devices (LVADs) prolong life and improve quality of life in patients with end-stage heart failure, therefore they are not strictly a palliative therapy.
- Risk prediction tools for estimating survival after LVAD implant have limited accuracy.
- Complications in LVAD recipients are frequent and difficult to predict. The most significant complications are gastrointestinal bleeding, driveline/device infection, arrhythmias, stroke and neurologic events, right ventricular failure, and device failure.
- Symptom management for LVAD complications is an important aspect of LVAD care.
- Palliative care consultation is appropriate when LVAD therapy is initially considered, when complications arise, and at the end of life.
- Identification of patient perceptions of life with specific LVAD complications will help improve delivery of care to LVAD recipients.
- LVAD deactivation is a reasonable choice at the end of life when patients are dying of LVAD complications or comorbid illness.

Care of Patients Dying with Advanced Heart Failure

Chronic heart failure affects 2.4 % of the adult population of the United States [1] and will further increase in prevalence as the population ages [2]. Estimates of 'end-stage' heart failure (ACC/AHA Stage D or NYHA Class IV) prevalence in the United States range from 0.2 % [3] to 5 % [4] of heart failure patients, with up to 300,000 patients affected. Heart failure at its end-stage follows a natural history of worsening exertion tolerance, worsening end-organ perfusion, and frequent hospitalizations accompanied by impaired cognition, anorexia, and other undesirable symptoms. As the burden of symptoms mounts, depression is increasingly common and the impact on patient quality of life is amongst the worst of any illness [5]. Beyond physical suffering, caregiver relationships and financial security may also be strained. For those failing conventional heart failure therapies, infusion of intravenous inotropes may improve cardiac output and end-organ perfusion, but it is unclear if inotropes improve overall quality of life [6], and lifespan clearly does not improve [7]. Therefore, in patients with symptomatic advanced heart failure refractory to conventional therapy, assessment of patient wishes for symptomatic vs. life-prolonging treatment is essential, and ACC/AHA guidelines confer a Class 1 recommendation for palliative care consultation in these patients [8].

While the natural history of end-stage heart failure is grim, there are two life-prolonging therapies for patients with advanced heart failure refractory to medical treatment – cardiac transplantation and left ventricular assist device (LVAD) implantation. Cardiac transplantation replaces the failing heart with a healthy donor heart,

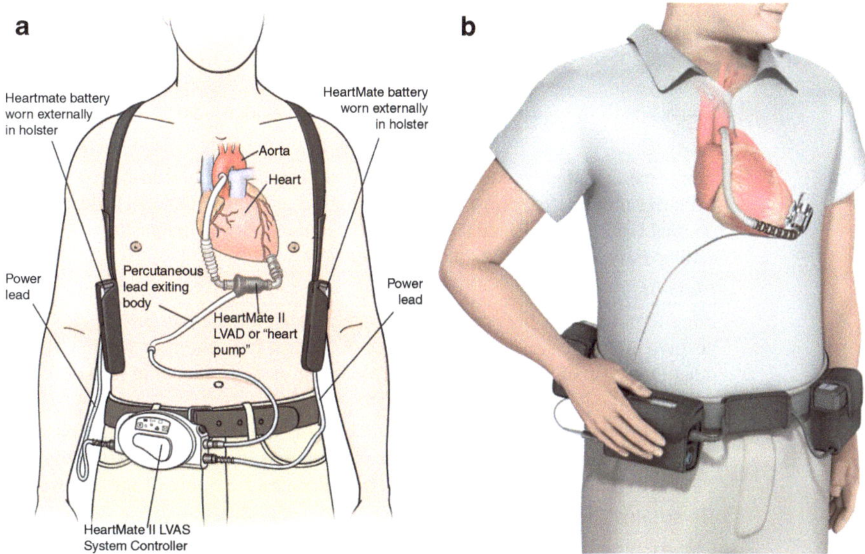

Fig. 13.1 FDA-approved permanent left ventricular assist devices (LVADs). (**a**) HeartMate II (Thoratec, Pleasanton, CA) (Reprinted with the permission of Thoratec Corporation). (**b**) HeartWare (HeartWare, Framingham, MA) (Reprinted with the permission of HeartWare, Inc.)

confers a 1-year survival rate of nearly 90 % and a median survival of 11 years, and improves quality of life [9]. However, owing to factors such as advanced age and comorbid illness, many patients with end-stage heart failure are ineligible for transplant. Even for those who are listed for transplant, organ availability is limited, with less than 2,000 transplants annually in the U.S. over the last decade [9]. LVADs are partially-externalized, surgically-implanted mechanical pumps that augment cardiac output and unload the failing left ventricle by propelling blood from the left ventricular apex and ejecting it into the aorta (Fig. 13.1). Survival is improved compared with medical therapy in both LVAD-supported patients awaiting transplant (bridge to transplant or BTT) and in those ineligible for transplant (destination therapy or DT). Unlike heart transplants, LVADs are not a limited resource. While heart transplantation has served an epidemiologically small but important role in the overall landscape of end-stage heart failure, LVADs by virtue of broader availability are reshaping the approach to managing end-stage systolic heart failure. Patients with end-stage heart failure refractory to medications should be evaluated for candidacy for both of these life-prolonging therapies [10]. For those in whom LVAD or transplant is not advisable or desired, guidelines recommend a palliative care approach [8]. However, for those patients who receive LVAD support and subsequently experience impaired quality of life or limited survival, the role and timing of palliative care is relatively undefined, as few studies have addressed these issues [11]. Major societies have general recommendations for the incorporation of palliative care into the management of patients with advanced heart failure, with or without LVAD support (Table 13.1).

Table 13.1 Guideline recommendations for palliative care/supportive care

ACC/AHA	*Class I*
	Palliative and supportive care is effective for patients with symptomatic advanced HF to improve quality of life (Level of evidence B)
	Class IIb
	Long-term, continuous intravenous inotropic support may be considered as palliative therapy for symptom control in select patients with stage D despite optimal GDMT and device therapy who are not eligible for either MCSD or cardiac transplantation (Level of evidence B)
	Class III
	Long-term use of either continuous or intermittent, intravenous parenteral positive inotropic agents, in the absence of specific indications or for reasons other than palliative care, is potentially harmful in the patient with HF (Level of evidence B)
HFSA	End-of-life care should be considered in patients who have advanced, persistent HF with symptoms at rest despite repeated attempts to optimize pharmacologic, cardiac device, and other therapies, as evidenced by 1 or more of the following:
	HF hospitalization (Strength of evidence B)
	Chronic poor quality of life with minimal or no ability to accomplish activities of daily living (Strength of evidence C)
	Need for continuous intravenous inotropic therapy support (Strength of evidence B)
ESC	Patients in whom palliative care should be considered:
	Frequent admission to hospital or other serious episodes of decompensation despite optimized treatment
	Heart transplantation and mechanical circulatory support ruled out
	Chronic poor quality of life with NYHA class IV symptoms
	Cardiac cachexia/low serum albumin
	Dependence in most activities of daily living
	Clinically judged to be close to the end of life
ISHLT	*Class I*:
	Consultation with palliative medicine should be considered prior to MCSD implantation to facilitate discussion of end of life issues and establish an advance directive or living will, particularly when implanted as DT. (Level of evidence C)
	In situations when there is no consensus about discontinuing MCSD support, consideration may be given to consulting with the hospital ethicist or ethics board. (Level of evidence C)
	A multidisciplinary team lead cooperatively by cardiac surgeons and cardiologists, and composed of subspecialists (i.e., palliative care, psychiatry, and others as needed), MCSD coordinators, and other ancillary specialties (i.e., social worker, psychologist, pharmacist, dietitian, physical therapist, occupational therapist, and rehabilitation services) is indicated for the in-hospital management of MCSD patients. (Level of evidence C).

(continued)

Table 13.1 (continued)

	Class IIa
	Palliative care consultation should be a component of the treatment of end-stage heart failure, and it should be addressed during the evaluation phase for MCSD. In addition to symptom management, goals and preferences for end of life should be discussed with patients receiving MCSD as DT. (Level of evidence C)

ACC/AHA: American College of Cardiology/American Heart Association Guideline for the Management of Heart Failure 2013 [8]
HFSA: Heart Failure Society of America 2010 Comprehensive Heart Failure Practice Guideline [12]
ESC: European Society of Cardiology Guidelines for the diagnosis and treatment of acute and chronic heart failure 2012 [13]
ISHLT: International Society of Heart and Lung Transplantation Mechanical Circulatory Support Guidelines 2013 [10]
DT destination therapy, *GDMT* guideline-directed medical therapy, *HF* heart failure, *MCSD* mechanical circulatory support device, *NYHA* New York Heart Association

The Impact of Left Ventricular Assist Devices on the Course of Advanced Heart Failure

Durable LVADs were approved by the United States FDA as a bridge to heart transplant in 1994 and later approved for destination therapy in 2003. Earlier generation pulsatile devices were prone to device malfunction and carried a high morbidity, yet in the REMATCH randomized trial 1-year survival was 52 % with an LVAD compared to 23 % on optimal medical therapy [14]. Pulsatile LVADs have been supplanted by continuous-flow devices, most notably the HeartMate II (Thoratec, Pleasanton, CA, approved for BTT and DT) and the HeartWare LVADs (HeartWare, Framingham, MA, approved for BTT). These devices involve either axial (HeartMate II) or centrifugal (HeartWare) mechanisms to continuously propel blood from the left ventricular apex to the ascending aorta. Compared to pulsatile devices, these newer LVADs are smaller, more durable, and easier to implant. They offer improved survival compared to previous generation devices in patients being listed for heart transplant and in patients ineligible for transplantation [15, 16]. The most recent data from the Interagency Registry for Mechanically Assisted Circulatory Support (INTERMACS) demonstrate an actuarial 2 year survival of approximately 60 % in DT continuous flow LVAD recipients and 70 % in BTT recipients [17]. No contemporary cohort of medically-treated patients provides an adequate comparison, but previous trial populations with NYHA Class IV heart failure treated medically experienced a 2 year survival of just 12 % [14]. Recipients of a LVAD, whether for BTT or DT, experience improvements in heart failure symptom scores, overall health-related quality of life, and mobility [18]. Based on data combined from trials of the HeartMate II, mean New York Heart Association (NYHA) functional status improved from class IV pre-LVAD to class II post-LVAD. Since LVADs improve the quality of life *and* significantly prolong life, they are not a strictly palliative therapy. However, as discussed below, life after LVAD implantation may present challenges that require supportive or palliative approaches to maximize patient quality of life and well-being.

Complications of Left Ventricular Assist Device Support and Management

Despite improvements in survival, heart-failure specific and overall quality of life, and organ perfusion and function, the burden of LVAD-associated complications is considerable and difficult to predict. Major complications include infection, bleeding, device malfunction, stroke, and death. By 6 months post-implant, 60 % of patients have experienced at least one of these events and by 2 years 80 % have experienced at least one event [17]. Some events, e.g. driveline infection, may be readily controlled and may not recur, but in some patients the burden of device-related complications can be large. An individualized approach to management is employed, but there is much work to be done to determine complication-specific patient perceptions of quality of life. Managing LVAD complications requires evaluation of current quality of life, the potential reversibility of the complication, the burden of therapies to treat the complication, and patient expectations for future quality of life. If complications progress, worsen in severity, or increase in frequency, it is necessary to re-evaluate goals of care. Some patients may desire a palliative approach, particularly those who are no longer eligible for transplant. Table 13.2 summarizes the frequency of occurrence and the therapeutic approach to common LVAD complications, and these are discussed in more detail below.

Right Ventricular Failure

Right ventricular failure (RVF) occurs at a rate of approximately 20 % with continuous-flow LVADs [17] and may eventually occur in up to 25 % of supported patients [22]. Patients with early post-LVAD RVF have higher mortality at 1 year [19] and impaired exercise tolerance [23], although the impact of RVF on overall quality of life is not well-described. These patients manifest predominantly 'right-sided' symptoms of heart failure with abdominal fullness, anorexia, and peripheral edema with relatively little orthopnea unless ascites or pleural fluid accumulations cause a reduction in lung capacity. In the perioperative period, right-ventricular assist device implantation may be required, prolonging the hospital course [19]. For those with symptomatic congestion, diuretics are the mainstay of therapy. In patients with concomitant pulmonary hypertension, oral pulmonary vasodilators, such as sildenafil, may have a role in unloading the right ventricle and reducing central venous pressures [24]. Digoxin may improve acute hemodynamics in pulmonary hypertension [25], but there are no data to support its use in the management of RV failure associated with LVADs. If symptoms are refractory to oral therapies,

Table 13.2 Complications of left ventricular assist device support and management[a]

Complication	Frequency (per 100 patient-months)	Timeframe	Conditional survival	Management
Right ventricular failure	1.8	First 1–2 months greatest hazard	Survival 59 % at 1 year vs. 79 % in unaffected [19]	Diuretics (esp. torsemide) Pump speed reduction Intravenous inotropes Pulmonary vasodilators
Gastrointestinal bleeding	9.5[b]	Days to years	Similar to non-bleeders	Reversal of anticoagulation for severe bleeding Reduced antiplatelet therapy Reduced INR goals Adjunctive therapies: octreotide, oral contraceptives, thalidomide Pump speed reduction Gastrointestinal consultation for endoscopy/intervention
Epistaxis	–			Nasal packing Secondary prevention – saline spray, lubricant ENT consultation – cauterization
Device malfunction	1.6	Constant event rate	1 year survival 70 % after exchange	Surgical replacement of pump Intravenous inotropes Supportive transfusion for associated hemolysis
Stroke	1.8	Rate: 7 % at 6 months, 11 % at 12 months, 17 % at 24 months	Unclear	Physical/occupational/speech therapy Anticoagulation modifications Inspect for device thrombosis
Infection	8	Constant event rate ~15 % per year	Reduced survival if septicemic [20]	IV antibiotics transitioned to chronic oral antibiotics Driveline revision, surgical debridement
Ventricular Arrhythmia	4.7	~30 % incidence within 1st year, most evident within first month [21]	No clear difference in survival	Antiarrhythmic drug therapy ICD reprogramming to minimize shocks Evaluation for suction – speed reduction Radiofrequency ablation

(continued)

Table 13.2 (continued)

Complication	Frequency (per 100 patient-months)	Timeframe	Conditional survival	Management
Renal failure	1.4	Generally improves post-LVAD	Mortality 50 % at 1 year if remaining on dialysis	Evaluate cardiac output, nephrotoxic insults
				Hemodialysis

[a]As reported in 5th INTERMACS Annual Report unless otherwise noted [17]
[b]All causes of bleeding

intravenous inotropes may be required. Pump speed may be reduced to decrease RV workload. Patients who are awaiting heart transplant on mechanical right ventricular support or inotropic support generally require hospitalization for medical management and physical therapy in order to maintain adequate conditioning. Patients ineligible for heart transplant may benefit symptomatically from continuous intravenous inotropic support and palliative care consultation should be considered.

Bleeding Complications

Bleeding complications occur frequently on LVAD support, and the most common types of bleeding are gastrointestinal (GI) hemorrhage and epistaxis. Single-center data suggest that GI bleeding occurs in about 20 % of LVAD patients [26]. While some GI bleeding is due to gastritis or hemorrhoids, patients with continuous flow LVADs are subject to GI bleeding related to the formation of superficial arteriovenous malformations in the intestinal mucosa. Additionally, these patients develop an acquired von-Willebrand syndrome due to lysis of von-Willebrand factor multimers, resulting in an impairment of primary hemostasis [27, 28]. LVAD patients presenting with significant GI bleeding generally have anticoagulation held or reversed and often undergo extensive evaluation to determine the source of bleeding. This may include upper GI endoscopy, colonoscopy, capsule endoscopy or, in the case of brisk bleeding, tagged red blood cell nuclear scan or angiography. Hospitalizations are often protracted as anticoagulation is resumed and the patient observed for recurrence of bleeding. In patients who have had more than one bleeding episode, anticoagulation is generally reduced, with a lower dose of aspirin and/or lower target INR. While this may decrease the likelihood of further bleeding, it may also increase the likelihood of pump thrombosis, resulting in pump dysfunction or embolic complications. Occasionally GI bleeding is refractory and anticoagulation must be discontinued altogether. Patients may also be placed on adjunctive therapies such as thalidomide (which reduces vascular endothelial growth factor expression), subcutaneous octreotide (which reduces gastric acidity and induces portal and splanchnic vasoconstriction), or oral estrogen (which is prothrombotic). Data supporting these adjuvant strategies are limited and further investigation is needed. Despite these measures, some patients have low-grade chronic or intermittent GI blood loss and require periodic transfusions to maintain control of symptoms. Because less pulsatile flow is associated with greater degrees of GI bleeding [29] and discontinuation of continuous flow may 'cure' GI bleeding, patients may be managed with reduced LVAD support at lower pump speed.

Epistaxis is another potential complication of LVAD support and may share some of the same causal mechanisms as GI bleeding. Severe bleeding may require hospitalization with blood transfusion and nasal packing. Patients should be instructed to keep the nares moist with nasal saline or lubricant gel and to use oxymetazoline at the onset of bleeding episodes. Antiplatelet and anticoagulant management strategies are similar to those employed for GI bleeding.

Device Malfunction

Device malfunction or pump failure occurs in approximately 10 % of patients during the first 2 years of support with continuous flow pumps, leading to pump exchange or death [17]. Clinically, pump dysfunction is usually discovered after the emergence of heart failure symptoms and/or intravascular hemolysis, which is caused by shearing of red blood cells as they traverse a region of pump thrombosis. Pump replacement surgery may be performed with moderate morbidity and a 1-year survival rate post-exchange of 70 % [30]. The alternative to pump exchange for those patients unwilling to undergo further surgery is inotropic therapy and supportive care. These patients may require blood transfusions to maintain acceptable hemoglobin concentrations and severe hemolysis may induce arterial vasospasm through a mechanism of nitric oxide depletion resulting in visceral pain. For acute pump dysfunction, a strategy of thrombolysis or augmented anticoagulation (e.g. glycoprotein IIb/IIIa inhibitors) has been employed in an attempt to resolve thrombosis. Despite some reported successes [31, 32], it is our experience that the risk for major bleeding is high and likely prohibitive. As patients weigh the uncertainty of further surgery against mortality and impaired quality of life, palliative care consultation is particularly helpful for clarifying patient and family goals.

Neurologic Events

Neurologic events are the most feared complication of LVAD support and occur at a cumulative rate of 17 % at 2 years [17]. Strokes may be ischemic, from device-related emboli, or hemorrhagic resulting from anticoagulation. The severity of neurologic injury varies greatly. For the most severely affected patients, the burden on caregivers, already heavy in the routine care of an LVAD, may become unbearable. Patient and family goals of care must be readdressed in the wake of a major neurologic event. The overall prognosis after stroke in LVAD *per se* is not well-described and conventional post-stroke consultative services (speech, occupational, and physical therapy) are appropriate. Anticoagulation therapy may be interrupted and a decision to resume therapy must weigh concerns of neurologic bleeding with risk of thrombotic or embolic events.

Infection

Infectious complications occur frequently in continuous-flow LVADs, with approximately 20 % of patients experiencing a device-related infection within 1 year of implant [17]. The most frequent complication is infection of the LVAD driveline, which may range in presentation from increased drainage from the exit site at the skin to frank sepsis and deep tissue infection (Fig. 13.1). Extensive infection of the driveline may require surgical debridement with repositioning in healthier tissue. Infection of the pump itself may require revision of the pump pocket. Patients with septicemia have reduced survival compared with uninfected patients, but survival does not appear to be reduced in patients with localized infections without septicemia [20]. Early aggressive diagnostic workup accompanied by antibiotic therapy and the assistance of infectious disease consultation is warranted. Infection may become chronic but manageable with oral antibiotics, although the emergence of antimicrobial resistance is a concern.

Arrhythmia

Ventricular arrhythmias are frequent in patients on LVAD support, with nearly 1/3 of patients experiencing ICD shocks within the first year of support [21]. While LVADs reduce the overall burden of ventricular arrhythmias, with smaller LV cavity size, suction of the endocardium into the LVAD inflow cannula may cause ventricular arrhythmias. In patients experiencing ventricular arrhythmias, evaluation of LV size and cannula positioning by echocardiography is necessary. LVADs are also able to register suction events in a retrievable electronic log that may be reviewed by clinicians. LV cavity under-filling can be improved by discontinuing diuretics, increasing hydration, or reducing the pump speed. Most LVAD recipients already have an ICD and the presence of an ICD may be associated with improved survival [33]. However, ICD shocks are psychologically disturbing and may result in significant post-traumatic stress [34]. Further, patients who receive shocks may experience reduced survival [35]. ICD device programming should minimize shocks in most LVAD patients, with liberal use of anti-tachycardia pacing and defibrillation reserved for heart rates likely to result in hemodynamic instability (e.g. over 200 beats/min). LVAD recipients may tolerate prolonged ventricular arrhythmias because of the hemodynamic support afforded by the LVAD, but prolonged ventricular tachycardia (VT) may lead to right heart failure and associated symptoms. For patients with refractory VT who are sufficiently stable, elective defibrillation in a controlled setting with anesthesia is preferred. To prevent ICD shocks, antiarrhythmic drugs such as amiodarone or mexilitine may be needed, but their associated toxicities, particularly neurologic and gastrointestinal side effects, should be weighed against their potential benefit. For those who have failed medications, radiofrequency ablation of VT foci may be effective [36, 37]. Finally, in the care of the dying patient with an LVAD and refractory ventricular arrhythmia, turning off the ICD to prevent the pain and distress of shocks should be considered.

Renal Failure

Renal function improves initially after LVAD implantation due to improved perfusion, but over time renal function may deteriorate [38]. For patients who proceed to end-stage renal disease, hemodialysis is complicated by a relative lack of pulsatile blood flow and difficulties in measuring blood pressure, leading to under-dialysis and chronic congestion in some cases. Consequently, few dialysis centers are able to accommodate LVAD patients. Furthermore, renal failure may occur with other end-organ dysfunction at the end of life in patients failing LVAD support. Some patients elect not to pursue chronic dialysis. For those who do, survival is reduced, with a 1-year survival of approximately 50 % [17]. For patients unlikely to proceed to heart-kidney transplant, honest discussion of prognosis is crucial, and palliative approaches to renal failure may be preferable to dialysis.

LVAD Complications and Implications for Transplant Status

Complications in LVAD-supported patients awaiting heart transplant are a double-edged sword. The United Network for Organ Sharing (UNOS) allows for the highest priority transplant listing, Status 1A, in patients on "mechanical circulatory support with objective medical evidence of significant device-related complications such as thromboembolism, device infection, mechanical failure, or life-threatening ventricular arrhythmias" [39]. However, major complications may result in de-listing of candidates who are no longer fit for transplantation, and contemporary data suggest that patients with LVADs who are upgraded to Status 1A for a complication are at greater hazard for both death (HR 1.47, 95 % CI 1.00–2.18) and death or delisting (HR 1.75, 95 % CI 1.26–2.42) compared with

LVAD-supported transplant candidates who do not have their status upgraded [40]. Navigating between the threat of de-listing or death and the promise of a greater likelihood for transplant can be emotionally taxing and an important opportunity for enlisting palliative care consultation.

Transition to Palliative Care

When patients have experienced a major complication or have reached a point where the cumulative morbidity has exceeded the perceived benefit of ongoing support, patients and their families may wish to discuss discontinuation of LVAD therapy. This is an emotionally delicate decision and ethics consultation may be warranted in some cases [10]. Arguments for deactivation center on patient autonomy, beneficence, and the weighing of benefits against burdens [41]. However, as highlighted by two reports on LVAD discontinuation, the decision to terminate support is most often made by a surrogate as a result of patient incapacity, and advance directives rarely offer specific guidance for such a situation [42, 43]. From a legal standpoint, most view LVAD discontinuation as the removal of a supportive therapy (like a mechanical ventilator), and we are unaware of any legal repercussions faced by discontinuation of LVAD support at the behest of the patient or family. Nevertheless, this challenge highlights the importance of proactive palliative care involvement, advance directives, and end-of-life planning with the patient and family prior to the onset of incapacity. Some patients or family may desire complete device removal; however, device explantation is highly morbid and rarely performed outside the context of normalized left ventricular systolic function. There may be a role for less morbid surgical approaches to discontinue LVAD support and exclude the pump from the circulation in patients with refractory GI bleeding, but these are not widely practiced. In most cases, patients facing the end of life from a comorbid condition (malignancy, renal failure, disabling stroke) will continue LVAD support until death. Discontinuation of anticoagulation and any medications not providing symptomatic benefit is reasonable in these patients. Palliative care consultation and a focus on symptomatic care is essential and elicitation of patient and family desires for end-of life care, such as dying at home or in hospital, is needed. The option of hospice services for dying LVAD patients is center-specific. As many hospice programs are not equipped to manage LVAD patients, hospice is often an option only for patients who have decided to discontinue LVAD support. Knowledge of local hospice policies is required. Whatever the setting, discontinuation of LVAD support is a potentially distressing process and controller alarms may disturb an otherwise peaceful death. The textbox below presents our institution's protocol for discontinuing LVAD support, with advice on minimizing alarms. While LVAD operations manuals do not contain these instructions, manufacturers can provide centers with support for device discontinuation.

Discontinuing Left Ventricular Assist Device Support

Background

1. Ensure family, patient, caregivers have anticipated patient needs
2. Consider having family practice on a mock device

Device-Specific Instructions

HeartMate II (Thoratec, Pleasanton, CA)

1. Disconnect driveline from controller – will cause a red heart alarm.
2. At driveline connection to controller, turn Perc Lock to unlocked position.
3. Press down metal tab, pull out driveline.
4. Disconnect power from controller – red heart alarm will stop

HeartWare VAS (HeartWare, Framingham, MA)

1. Take red alarm adapter from patient's back-up controller and plug it into blue port of primary controller. If the red alarm adapter is not attached to controller before removing driveline and power sources, controller will continue to alarm.
2. Disconnect driveline from primary controller – controller will alarm
3. Pull back gray driveline cover until metal connection is seen.
4. Disconnect driveline by pulling metal connector away from controller.
5. Disconnect both power sources from controller – alarm will stop.

Integrated Approach, Patient Selection, Preparedness Planning

The ideal scenario for care in patients with advanced heart failure being considered for LVAD support would include accurate prediction of survival with and without LVAD, accurate prediction of LVAD-specific complications, knowledge of patient values regarding living with these complications, and overall patient expectations for life after LVAD implantation. However, there are practical obstacles to consider. First, near-term mortality in ambulatory patients with advanced heart failure is imperfectly predicted with current risk tools such as the Seattle Heart Failure Model and the Heart Failure Survival Score. Risk models for survival post-LVAD, such as the HeartMate Risk Score, are even less reliable. Second, more than half of patients with LVADs are clinically unstable at the time of implant and 16 % are in cardiogenic shock. As a result, surrogate decision-makers are often faced with the taxing proposition of deciding between LVAD support and a high likelihood of near-term death [17]. Pre-existing provider relationships can help, and prior clarification of patient wishes specifically regarding LVAD implantation is particularly useful. Third, even in patients who are counseled on potential LVAD complications, it is unclear whether adequate understanding of life with these unique complications can be conveyed. There is a role for mentorship by patients currently supported by LVADs for patients considering LVAD support, although a survivor bias clearly exists. A comprehensive model for management of patients supported by or being considered for an LVAD would include involvement of palliative services at key time-points from decision-making to late follow-up, as described by Goldstein et al. [11].

There remains a great deal of complexity and uncertainty in decision-making and management of these patients, and the expertise of palliative care specialists will continue to shape the optimal delivery of this evolving technology to those in need.

Role of Palliative Care Consultation at Key Stages in LVAD Care

Evaluation for LVAD Candidacy

Ensuring patient comprehension of treatment options, risks and benefits
Assist in eliciting patient preferences for care
Identifying lifestyle changes to be faced after LVAD implant
Identifying caregiver changes to be faced after LVAD implant
Advance directives, selection of healthcare power of attorney

Support After LVAD Implant

Dealing with complications of LVAD support
Reassessing patient goals of care and eliciting preferences after complications
Symptom management
Dealing with patients becoming ineligible for transplant after complications

At the End of Life

Ensuring patient and family comprehension of prognosis and goals of care
Symptom management
Family/caregiver support for dealing with end-of-life planning
Coordinating LVAD discontinuation

Data from Goldstein et al. [11]
LVAD left ventricular assist device

References

1. Roger VL, Go AS, Lloyd-Jones DM, et al. Heart disease and stroke statistics–2012 update: a report from the American Heart Association. Circulation. 2012;125(1):e2–220.
2. Heidenreich PA, Trogdon JG, Khavjou OA, et al. Forecasting the future of cardiovascular disease in the United States: a policy statement from the American Heart Association. Circulation. 2011;123(8):933–44.
3. Ammar KA, Jacobsen SJ, Mahoney DW, et al. Prevalence and prognostic significance of heart failure stages: application of the American College of Cardiology/American Heart Association heart failure staging criteria in the community. Circulation. 2007;115(12):1563–70.
4. Adler ED, Goldfinger JZ, Kalman J, Park ME, Meier DE. Palliative care in the treatment of advanced heart failure. Circulation. 2009;120(25):2597–606.
5. Bekelman DB, Havranek EP, Becker DM, et al. Symptoms, depression, and quality of life in patients with heart failure. J Card Fail. 2007;13:643–8.
6. Felker GM, O'Connor CM. Inotropic therapy for heart failure: an evidence-based approach. Am Heart J. 2001;142(3):393–401.
7. Stevenson LW. Clinical use of inotropic therapy for heart failure: looking backward or forward? Part II: chronic inotropic therapy. Circulation. 2003;108(4):492–7.
8. Yancy CW, Jessup M, Bozkurt B, et al. 2013 ACCF/AHA guideline for the management of heart failure: a report of the American College of Cardiology Foundation/American Heart Association Task Force on Practice Guidelines. J Am Coll Cardiol. 2013;62:e147–239.
9. Stehlik J, Edwards LB, Kucheryavaya AY, et al. The Registry of the International Society for Heart and Lung Transplantation: 29th official adult heart transplant report–2012. J Heart Lung Transplant. 2012;31(10):1052–64.
10. Feldman D, Pamboukian SV, Teuteberg JJ, et al. The 2013 International Society for Heart and Lung Transplantation Guidelines for mechanical circulatory support: executive summary. J Heart Lung Transplant. 2013;32(2):157–87.
11. Goldstein NE, May CW, Meier DE. Comprehensive care for mechanical circulatory support: a new frontier for synergy with palliative care. Circ Heart Fail. 2011;4(4):519–27.
12. Heart Failure Society of America. HFSA 2010 comprehensive heart failure practice guideline. J Card Fail. 2010;16(6):e1–192.
13. McMurray JJ, et al. ESC Guidelines for the diagnosis and treatment of acute and chronic heart failure 2012: the Task Force for the Diagnosis and Treatment of Acute and Chronic Heart Failure 2012 of the European Society of Cardiology. Developed in collaboration with the Heart Failure Association (HFA) of the ESC. Eur Heart J. 2012;33(14):1787–847.
14. Rose E, et al. Long-term use of a left ventricular assist device. N Engl J Med. 2001;345(20):1435–43.
15. Miller LW, Pagani FD, Russell SD, et al. Use of a continuous-flow device in patients awaiting heart transplantation. N Engl J Med. 2007;357(9):885–96.
16. Slaughter MS, Rogers JG, Milano CA, et al. Advanced heart failure treated with continuous-flow left ventricular assist device. N Engl J Med. 2009;361(23):2241–51.
17. Kirklin JK, Naftel DC, Kormos RL, et al. Fifth INTERMACS annual report: risk factor analysis from more than 6,000 mechanical circulatory support patients. J Heart Lung Transplant. 2013;32(2):141–56.
18. Rogers JG, Aaronson KD, Boyle AJ, et al. Continuous flow left ventricular assist device improves functional capacity and quality of life of advanced heart failure patients. J Am Coll Cardiol. 2010;55(17):1826–34.
19. Kormos RL, Teuteberg JJ, Pagani FD, et al. Right ventricular failure in patients with the HeartMate II continuous-flow left ventricular assist device: incidence, risk factors, and effect on outcomes. J Thorac Cardiovasc Surg. 2010;139(5):1316–24.
20. Topkara VK, Kondareddy S, Malik F, et al. Infectious complications in patients with left ventricular assist device: etiology and outcomes in the continuous-flow era. Ann Thorac Surg. 2010;90(4):1270–7.

21. Oswald H, Schultz-Wildelau C, Gardiwal A, et al. Implantable defibrillator therapy for ventricular tachyarrhythmia in left ventricular assist device patients. Eur J Heart Fail. 2010;12(6):593–9.
22. Santambrogio L, Bianchi T, Fuardo M, et al. Right ventricular failure after left ventricular assist device insertion: preoperative risk factors. Interact Cardiovasc Thorac Surg. 2006;5(4):379–82.
23. Hasin T, Topilsky Y, Kremers WK, et al. Usefulness of the six-minute walk test after continuous axial flow left ventricular device implantation to predict survival. Am J Cardiol. 2012;110(9):1322–8.
24. Tedford RJ, Hemnes AR, Russell SD, et al. PDE5A inhibitor treatment of persistent pulmonary hypertension after mechanical circulatory support. Circ Heart Fail. 2008;1(4):213–9.
25. Rich S, Seidlitz M, Dodin E, et al. The short-term effects of digoxin in patients with right ventricular dysfunction from pulmonary hypertension. Chest. 1998;114(3):787–92.
26. Kushnir VM, Sharma S, Ewald GA, et al. Evaluation of GI bleeding after implantation of left ventricular assist device. Gastrointest Endosc. 2012;75(5):973–9.
27. Uriel N, Pak S-W, Jorde UP, et al. Acquired von Willebrand syndrome after continuous-flow mechanical device support contributes to a high prevalence of bleeding during long-term support and at the time of transplantation. J Am Coll Cardiol. 2010;56(15):1207–13. Cited 12 Jul 2013.
28. Crow S, Chen D, Milano C, et al. Acquired von Willebrand syndrome in continuous-flow ventricular assist device recipients. Ann Thorac Surg. 2010;90(4):1263–9; discussion 1269.
29. Wever-Pinzon O, Selzman CH, Drakos SG, et al. Pulsatility and the risk of non-surgical bleeding in patients supported with the continuous-flow left ventricular assist device HeartMate II. Circ Heart Fail. 2013;6(3):517–26. Available from: http://www.ncbi.nlm.nih.gov/pubmed/23479562. Cited 22 May 2013.
30. Moazami N, Milano CA, John R, et al. Pump replacement for left ventricular assist device failure can be done safely and is associated with low mortality. Ann Thorac Surg. 2013;95(2):500–5.
31. Al-Quthami AH, Jumean M, Kociol R, et al. Eptifibatide for the treatment of HeartMate II left ventricular assist device thrombosis. Circ Heart Fail. 2012;5(4):e68–70.
32. Thomas MD, Wood C, Lovett M, Dembo L, O'Driscoll G. Successful treatment of rotary pump thrombus with the glycoprotein IIb/IIIa inhibitor tirofiban. J Heart Lung Transplant. 2008;27(8):925–7.
33. Refaat MM, Tanaka T, Kormos RL, et al. Survival benefit of implantable cardioverter-defibrillators in left ventricular assist device-supported heart failure patients. J Card Fail. 2012;18(2):140–5.
34. Sears SF, Hauf JD, Kirian K, Hazelton G, Conti JB. Posttraumatic stress and the implantable cardioverter-defibrillator patient: what the electrophysiologist needs to know. Circ Arrhythm Electrophysiol. 2011;4(2):242–50.
35. Ambardekar AV, Allen LA, Lindenfeld J, et al. Implantable cardioverter-defibrillator shocks in patients with a left ventricular assist device. J Heart Lung Transplant. 2010;29(7):771–6.
36. Osaki S, Alberte C, Murray MA, et al. Successful radiofrequency ablation therapy for intractable ventricular tachycardia with a ventricular assist device. J Heart Lung Transplant. 2008;27(3):353–6.
37. Dandamudi G, Ghumman WS, Das MK, Miller JM. Endocardial catheter ablation of ventricular tachycardia in patients with ventricular assist devices. Heart Rhythm. 2007;4(9):1165–9.
38. Hasin T, Topilsky Y, Schirger JA, et al. Changes in renal function after implantation of continuous-flow left ventricular assist devices. J Am Coll Cardiol. 2012;59(1):26–36.
39. OPTN Policy 3.7 – Allocation of Thoracic Organs. 2013:1–45. http://optn.transplant.hrsa.gov/PoliciesandBylaws2/policies/pdfs/policy_9.pdf.
40. Wever-Pinzon O, Drakos SG, Kfoury AG, et al. Morbidity and mortality in heart transplant candidates supported with mechanical circulatory support: is reappraisal of the current United network for organ sharing thoracic organ allocation policy justified? Circulation. 2013;127(4):452–62.

41. Lachman VD. Left ventricular assist device deactivation: ethical issues. Medsurg Nurs. 2011;20(2):98–100.
42. MacIver J, Ross HJ. Withdrawal of ventricular assist device support. J Palliat Care. 2005;21(3):151–6.
43. Mueller PS, Swetz KM, Freeman MR, et al. Ethical analysis of withdrawing ventricular assist device support. Mayo Clin Proc. 2010;85(9):791–7.

Chapter 14
Assessment and Management of Cognitive Dysfunction and Frailty at End of Life

Jonathan Afilalo and Caroline Michel

Abstract One out of two patients with advanced cardiovascular disease exhibits signs of frailty or cognitive impairment. Identification of frailty or cognitive impairment, when objectified with a validated instrument, confers a negative effect on prognosis and may be an important clue that the patient with advanced cardiovascular disease is entering the end-of-life phase. These patients have special needs that are best addressed in conjunction with a multidisciplinary palliative care team. Challenges include recognizing and managing pain and other symptoms, assessing competence to make medical decisions, maintaining informed communication with the patients and their families, and differentiating potentially treatable etiologies such as: low cardiac output with hypoperfusion, sleep-disordered breathing, metabolic disturbances, delirium, and depression. Advanced care planning should be encouraged, with the foreseeable outcomes framed around the patient's cardiovascular disease status but also around their resiliency and homeostatic reserve. Ultimately, an approach of shared decision-making is central to the care of frail and cognitively impaired patients at the end-of-life.

Keywords Frailty • Cognitive impairment • Dementia • End of life • Heart failure • Cardiovascular disease

J. Afilalo, MD, MSc, FACC, FRCPC (✉) • C. Michel, MD, FRCPC
Division of Cardiology, McGill University,
3755 Cote Ste Catherine Rd, E-206,
Montreal, QC, H3T 1E2, Canada
e-mail: jonathan.afilalo@mcgill.ca

S.J. Goodlin, M.W. Rich (eds.), *End-of-Life Care in Cardiovascular Disease*,
DOI 10.1007/978-1-4471-6521-7_14

Key Points

- Frailty and cognitive impairment are highly prevalent to assess frailty status and cognitive function near the end-of-life
- An objective frailty assessment tool should be used rather than subjective judgment
- The Montreal Cognitive Assessment is the preferred tool to assess cognitive impairment in patients with advanced cardiovascular disease
- The utility of frailty and cognitive assessment is twofold: (i) to refine estimates of life expectancy and better identify those that are approaching end-of-life, (ii) to identify the subset of patients that may benefit most from palliative care services
- One must consider the broad differential diagnosis for cognitive impairment and frailty, which includes a number of potentially reversible causes
- Shared decision making, encompassing advanced care directives is essential

Prevalence of Frailty and Cognitive Impairment in Advanced Cardiovascular Disease

Frailty and cognitive impairment (CI) are pervasive, albeit not always appreciated, in patients with advanced cardiovascular disease (CVD) approaching end of life (EOL). The prevalence of frailty has been reported to be between 20 and 63 % in various cohorts of patients with advanced heart failure, severe coronary artery disease and valvular heart disease [1]. A similar prevalence of CI has been reported in advanced CVD, ranging between 25 and 74 %, with 10–20 % having severe CI [2–4]. The extent of CI further worsens when patients are in decompensated states [5]. Both are closely correlated with age; from 65 to 85, the prevalence of frailty and CI rises fivefold [6]. Since the elderly currently represent the fastest growing segment of our population, frailty and CI are sure to become increasingly frequent problems, particularly in older EOL patients.

Assessment Tools

Frailty

Frailty occurs as a result of clinical and subclinical losses in multi-organ function and reserve. Loss of muscle mass and strength, termed *sarcopenia*, is a pivotal aspect of frailty that is exacerbated in a perpetuating cycle by low physical activity, low metabolic rate, and malnutrition. Clinically, one may recognize and measure

Table 14.1 Fried frailty scale

Frailty domain	Method of measurement	Cutoffs for measurement		
1. Slowness	5-m gait speed *Patient is asked to walk at a comfortable pace from a 0-m start line to past a 5-m finish line, the cue to start and stop the stopwatch is the first footfall after the start line and first footfall after the finish line, this is repeated 3 times and the average time is recorded*	Sex- and height-based cutoff		
		♂	≤173 cm	**≤0.65 m/s**
			>173 cm	**≤0.76 m/s**
		♀	≤159 cm	**≤0.65 m/s**
			>159 cm	**≤0.76 m/s**
		Simplified alternative cutoff		
		♂/♀: **≤0.83 m/s**		
2. Weakness	Handgrip strength *Patient is asked to squeeze a handgrip dynamometer as hard as possible, this is repeated 3 times (with each hand and then with the strongest hand) and the maximum value is recorded*	Sex- and BSA-based cutoff		
		♂	≤24 kg/m^2	**≤29 kg**
			24.1–28 kg/m^2	**≤30 kg**
			>28 kg/m^2	**≤32 kg**
		♀	≤26 kg/m^2	**≤17 kg**
			26.1–29 kg/m^2	**≤18 kg**
			>29 kg/m^2	**≤21 kg**
		Simplified alternative cutoff		
		♂	**≤30 kg**	
		♀	**≤20 kg**	
3. Low physical activity	Minnesota Leisure-time physical activity questionnaire[103]	♂	**<383 kcal/week**	
		♀	**<270 kcal/week**	
4. Weight loss	Self-reported	**>10 lbs or >5 % over the past year**		
5. Exhaustion	2 questions: How often do you feel "Everything I did was an effort" "I could not get going"	Positive if answered either: "**Most of the time**" or "**Moderate amount of the time**"		
≥3 criteria required for a diagnosis of frailty				

From Fried et al. [7]. Reprinted with permission from Oxford University Press
BSA body surface area

the phenotype of frailty according to Fried's model of five criteria: slowness measured by 5-m gait speed test, weakness measured by handgrip strength test, shrinking measured by self-reported weight loss, exhaustion and inactivity measured by questionnaire; a score ≥3/5 signifies frailty (Table 14.1) [7]. Some advocate a more parsimonious model for frailty that includes solely 5-m gait speed, whereas other advocate a more expanded model that adds CI and mood disturbance as sixth and seventh criteria for frailty. An alternative approach to measuring frailty is Rockwood's model of up to 70 "deficits" (a diverse assortment of symptoms, conditions, functional items) that are summed in an individual patient [8].

At the physiological level, frailty is typified by impaired homeostatic reserve and reduced resiliency to stressors either related to illness or iatrogenic factors. When faced with exacerbations or invasive therapies, as is commonly encountered towards the EOL, frail patients tend to demonstrate marked and often disproportionate decompensation in health status and functional capacity. At the epidemiological level, frailty is strongly associated with risk of cardiac and all-cause mortality, complications after cardiovascular interventions, institutionalization, and lower quality of life [9, 10].

Disability is an inter-related but distinct concept from frailty, that is defined as inability or dependency to carry out basic and instrumental activities of daily living (ADL and IADL, respectively) [11]. The Katz [12], Barthel [13], and Older Americans Resources and Services (OARS) [14] scales are used in research and in practice to assess disabilities. Disabilities surface in the later stages of the frailty spectrum, once patients have faced significant stressors and accrued deficits. Importantly, disabilities in ADLs have been shown to be harbingers of EOL, signifying an expected survival less than 6 months in many independent studies [15, 16].

Cognitive Impairment

CI is defined as impairment in one or more cognitive function domains, namely: memory and learning, language, executive function, complex attention, perceptual-motor, and social cognition. Following the criteria set forth by the Diagnostic and Statistical Manual (DSM), these impairments must be acquired and represent a significant decline from baseline, they must interfere with independence in everyday activities, and not occur exclusively during the course of delirium. The diagnosis is insidious, and often missed or overlooked in practice. We and others have observed that dementia or mild CI is documented in admission notes and discharge summaries in 5–10 % of older cardiology ward patients, whereas it is elicited by cognitive testing in over one-third of such patients [17]. When CI is present but not documented or recognized by the treating physicians, the risk of mortality and hospital readmission increases significantly [2]. Mild CI may be falsely attributed to "normal aging" and even significant dementia may be attributed to behavioral or personality issues. This underscores the importance of cognitive testing rather than self-report or subjective assessment to establish the presence or absence of CI.

A number of cognitive testing instruments have been developed and validated. The Mini-Mental Status Examination (MMSE) [18] spans orientation, registration, attention and calculation, recall, and language. A score ≥27/30 is normal, 21–26 is mild impairment, 11–20 is moderate impairment, and ≤10 is severe impairment (another classification scheme states that ≥24 is normal, 18–23 is mild-moderate impairment, and ≤17 is severe impairment); cutoffs exist adjusted for age, sex, and educational level. The limitations of the MMSE are that it is less sensitive to mild CI, and it is proprietary. The Montreal Cognitive Assessment (MoCA) [19] – available at www.mocatest.org – overcomes these limitations, and furthermore, was found to be more predictive of adverse outcomes in heart failure patients [20, 21]. MoCA spans visuospatial and executive function, naming, memory, attention, language, abstraction, delayed recall, and orientation. A score <26 is abnormal; severity is determined by the burden of functional impairments. Brief instruments such as the Mini-Cog employ a clock drawing test and 3-word recall to efficiently screen for dementia (positive if 0 words recalled, or 1–2 words recalled plus abnormal clock draw).

Dementia may be sub-classified as neurodegenerative or non-neurodegenerative, with the former consisting of these major dementia syndromes: Alzheimer disease, vascular dementia, Lewy body disease, frontotemporal dementia, and Parkinson disease. Non-neurodegenerative etiologies must be carefully considered in patients with advanced CVD, since these tend to manifest in a subclinical yet nefarious fashion, consisting of medication side effects, metabolic disturbances, depression, and other differential diagnoses reviewed in section V.

Frailty and Cognitive Impairment as Cues to Consider Palliative Care

Frailty and CI Can Refine Predictions of Expected Survival

One of the primordial challenges in delivering effective EOL care is knowing when EOL is near in order to engage the patient in shared EOL decision making and activate palliative care (PC) resources in a timely manner. Patient and family satisfaction is increased when PC is initiated earlier in the disease trajectory. Prediction of expected survival is used by clinicians to guide this transition, which is gradual rather than abrupt, from curative to palliative therapies. Unfortunately, only 15 % of clinicians surveyed believed that they could reliably predict EOL in their heart failure patients [22], and risk models to predict expected survival in non-cancer conditions are inaccurate in the last 6 months of life. A systematic review of 11 risk models concluded that "no prognostic model can be recommended" [23]. Heart failure-specific risk models were not developed to predict 6-month survival, rather focusing on in-hospital (<30 days; ADHERE [24] and EFFECT [25] risk models) or long-term (>1–5 years; Seattle Heart Failure Model [26]) survival.

To address this, Huynh and Rich developed a risk model to predict 6-month survival tailored to older patients with heart failure who may be candidates for hospice care [27]. The following seven items emerged in multivariable analysis, with ≥4/7 items signifying a high risk of death at 6 months: age ≥75, dementia, coronary artery disease, peripheral arterial disease, systolic blood pressure <120 mmHg, BUN ≥30 mg/dL, Na <135 mEq/L. Dementia was among the most powerful predictors, with an adjusted hazard ratio of 2.02. In another study, the combination of MMSE score <18 plus Barthel ADL score <90 had a synergistic effect on predicting 6-month survival [28].

Among the risk models that predict 30-day survival, Pilotto and Ferrucci compared their Multidimensional Prognostic Index (MPI) risk score based on a comprehensive geriatric assessment including CI and ADL/IADL disabilities (Table 14.2 and Fig. 14.1) [29] to the ADHERE and EFFECT risk scores that are based solely on cardiac parameters and do not encompass geriatric parameters. The MPI substantially outperformed other risk scores (C-statistic 0.80–0.83 vs. 0.65–0.71) indicating the value of considering geriatric domains to predict EOL in vulnerable heart failure patients.

Table 14.2 Multidimensional prognostic index

	Problems		
	No	Minor	Severe
Assessment	(Value = 0)	(Value = 0.5)	(Value = 1)
ADL[a]	6–5	4–3	2–0
Instrumental ADL[a]	8–6	5–4	3–0
Short portable mental status questionnaire[b]	0–3	4–7	8–10
Comorbidity index[c]	0	1–2	≥3
Mini nutritional assessment[d]	≥24	17–23.5	<17
Exton-smith scale[e]	16–20	10–15	5–9
No. of medications	0–3	4–6	≥7
Social support network	Living with family	Institutionalized	Living alone

From Pilotto et al. [29]. Reprinted with permission from Wolters Kluwer Health
[a]No. of active functional activities
[b]No. of errors
[c]No. of diseases
[d]Mini Nutritional Assessment score: ≥24, satisfactory; 17–23.5, at risk; <17, malnutrition
[e]Exton-Smith Scale score: 16–20, minimum risk; 10–15, moderate risk; 5–9 high risk of developing pressure sores

Strictly speaking, frailty is not included in the aforementioned risk models, although there is evidence to suggest that frailty is strongly associated with 6-month mortality incremental to traditional risk models [30]. Frail older adults hospitalized with severe CAD had a two- to four-fold increase in 6-month mortality [31], lower QOL, lower physical and mental functioning scores [32] compared to their nonfrail counterparts. Similar data have been reported in advanced heart failure and valvular heart disease. Recurrent hospitalizations should alert the clinician that their patient is likely to be frail, and also likely to be approaching EOL.

Thus, it is abundantly clear that CI and frailty are powerful predictors of survival in CVD, and that integrating CI and frailty within the process of risk prediction refines estimates of survival, and allows clinicians to ascertain with greater confidence when EOL may be approaching in their patients.

Identify Patients That Benefit Most from Palliative Care

According to the revised framework, the indications for PC should be driven not by prognosis, but by the needs of the patients [33] (in comparison to hospice care, which is driven by expected survival <6 months). This extends beyond medical needs, to psychological and spiritual needs, as well as social and family-related needs. Such needs are preeminent in frail and CI patients. Frail patients have an

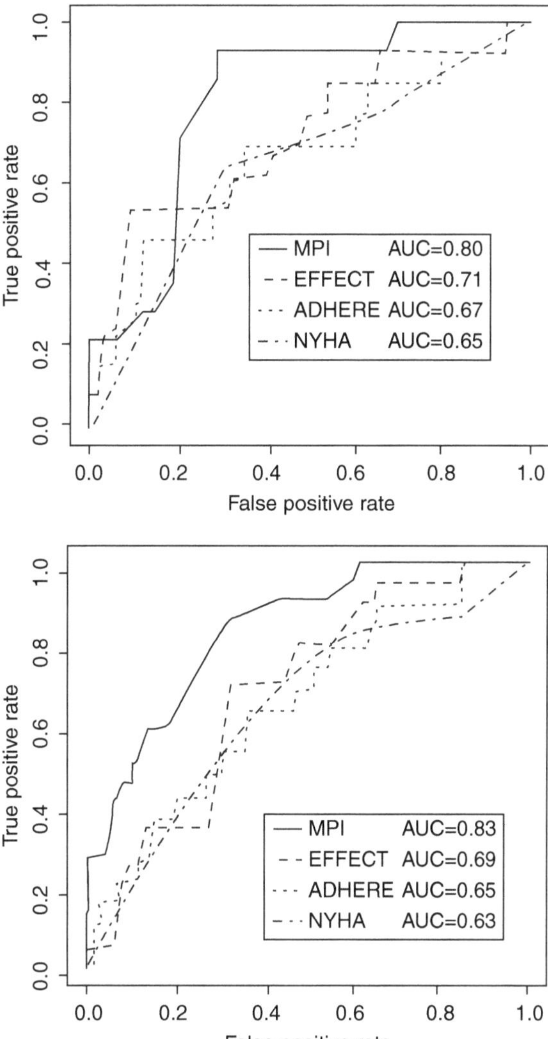

Fig. 14.1 ROC curves for the MPI, NYHA, EFFECT, and ADHERE risk scores at 30 days of follow-up in men (*bottom*) and women (*top*) (From Pilotto et al. [29]. Reprinted with permission from Wolters Kluwer Health)

increased frequency of depression, anxiety, and social isolation. CI patients have an increased frequency of untreated symptoms, poor self-care, and family stress [34]. The impact on family members may be severe, with a high incidence of depression and prolonged bereavement noted in those who were taking care of their relative with dementia [35]. These issues are optimally addressed and managed by a specialist in PC.

When comparing two EOL cohorts, one with end-stage heart failure and one with cancer, heart failure patients were less likely to receive a prescription for opiates (22 % vs. 46 %) and less likely to be referred for hospice care (20 % vs. 51 %) [36]. Heart failure patients were more likely to be admitted for acute care services in the

30 days preceding death (60 % vs. 45 %) and to be admitted to the intensive care unit for aggressive management (19 % vs. 7 %). CI was an independent predictor for not receiving an opiate prescription. Interestingly, another study that compared end-stage heart failure to cancer found that the needs for PC were equally justifiable in heart failure given the high symptom burden, depression, and low spiritual well-being [37].

Patients following the frailty and dementia trajectories have the highest risk of developing significant dependencies in ADLs towards the EOL [38, 39]. Active involvement of geriatrics, physiotherapy, occupational therapy, and home care services should be sought to limit the burden of disability and help patients maintain an independent or semi-independent lifestyle in the EOL period.

Different Diagnosis and Management of Frailty and Cognitive Impairment at End-of-Life

Approach to Cognitive Impairment in the EOL Cardiovascular Patient

An elderly patient presenting signs and symptoms of CI should not be assumed to have a form of neurodegenerative dementia, even if Alzheimers disease and vascular dementia are exceedingly common. In the setting of advanced CVD, a number of other potentially reversible causes should be sought (Fig. 14.2).

First, in the presence of a low cardiac output, cerebral hypoperfusion has been postulated to cause CI. The evidence to support this is mixed, with studies showing that cerebral blood flow by transcranial Doppler is [40] and is not [41] associated with CI, and studies showing that both hypotension [42] and hypertension [43] are associated with CI. Anecdotally, we have observed rapid and dramatic changes in cognitive function in some heart failure patients after a brief course of intravenous vasodilators or inotropic "holiday", although guidelines do not necessarily endorse this as routine practice. Second, embolic events due to occult atrial fibrillation or left ventricular thrombi should be excluded. Third, hypoxemia due to chronic heart failure and/or sleep apnea can cause CI, and extrapolating from the pulmonary literature, home oxygen therapy and nocturnal positive airway pressure have been shown to improve cognitive function [44, 45]. Fourth, metabolic disturbances such as uremic renal failure or severe hyponatremia may manifest as forgetfulness, confusion, and decreased attention span, and are usually improved by reducing diuretic dosage in prerenal failure or increasing diuretic dosage and restricting fluid intake in hyponatremia. Fifth, depression is one of the great masqueraders that must be excluded, as it is highly prevalent in patients with advanced CVD. Sixth, delirium should be differentiated from CI, especially in patients presenting with an acute change from baseline. Lastly, if CI is confirmed but there is no neurodegenerative disease or reversible cause identified, severe heart failure itself has been directly linked to CI by way of accelerated grey matter atrophy [46].

Fig. 14.2 Differential diagnosis of cognitive impairment in cardiac patients

Neurodegenerative diseases
- Alzheimer disease
- Vascular dementia (including embolic events from atrial fibrillation or mural thrombi)
- Lewy body disease
- Frontotemproral dementia
- Parkinson disease

Low cardiac output state

Chronic hypoxemia

Sleep apnea

Cardiorenal syndrome / uremia

Severe hyponatremia

Depression

Delirium

Heart failure-related grey matter atrophy

Approach to Frailty Signs and Symptoms in the EOL Cardiovascular Patient

Loss of muscle mass and strength are cardinal signs of frailty and also of advanced heart failure, termed sarcopenia in the former and cardiac cachexia in the latter [47]. Sarcopenia is a geriatric syndrome defined by low muscle mass and either low muscle strength or slow gait speed; its causes are multifactorial and may be equated to accelerated or unsuccessful aging of the musculature. Cachexia is a metabolic syndrome defined by weight loss and either low muscle mass, low muscle strength, fatigue, anorexia, hypoalbuminemia, anemia, or elevated inflammatory markers; it is caused by inflammatory cytokines associated with various pathological disease states. Evidently, there is room for overlap between these two syndromes, yet their pathophysiology and therapeutic options differ. Sarcopenia may respond to exercise and nutritional interventions and androgen replacement therapy [48], whereas cachexia is less often modifiable as it requires treatment of the underlying disease.

Anergia, exhaustion, and low physical activity are also cardinal symptoms of frailty with a broad differential diagnosis. Before attributing these symptoms to frailty, the clinician should exclude significant anemia, hypothyroidism, depression, sleep apnea, metabolic disturbances, iatrogenic, or worsening low cardiac output state. The presence of any of these should direct the clinician toward a targeted treatment, whereas if attributed to frailty, then the plan should revolve around a graded exercise intervention and perhaps a trial of cognitive behavioral therapy [49, 50].

Overcoming Barriers to Improve End-of-Life Care for the Frail and Cognitively Impaired

Predicting EOL

Clinicians are not equipped to predict EOL with an acceptable degree of accuracy using traditional risk models. The frailty and dementia trajectories toward death have been shown to be protracted and particularly difficult to predict. To improve risk prediction, clinicians are encouraged to integrate frailty and CI and/or utilize a model such as the MPI that encompasses geriatric domains rather than cardiac-only domains. Among frail patients, the severe form that is approaching EOL often has one or more of these features: muscle wasting, malnutrition, inability to mobilize, and disabilities in ADLs. Among CI patients, these features have been associated with <6-month survival: muscle wasting, anorexia, dehydration, pneumonia, and disabilities in ADLs [34].

Recognizing Symptoms

In advanced dementia, patients may have difficulty expressing symptoms of pain and dyspnea. Physicians and nurses express a lack of comfort and expertise to detect and quantify pain in CI patients, and benefit from continuing medical education in this regard. Moreover, instruments such as the Pain Assessment IN Advanced Dementia (PAINAD) scale (Table 14.3) [51] are available to assist in detecting pain by way of facial expressions, vocalization, and body language.

Table 14.3 Pain assessment in advanced dementia (PAINAD) scale

	0	1	2
Breathing (independent of vocalization)	Normal	Occasional labored breathing. Short period of hyperventilation	Noisy labored breathing. Long period of hyperventilation. Cheyne-stokes respirations
Negative vocalization	None	Occasional moan or groan. Low- level of speech with a negative or disapproving quality	Repeated troubled calling out. Loud moaning or groaning. Crying
Facial expression	Smiling or inexpressive	Sad, frightened, frown	Facial grimacing
Body language	Relaxed	Tense. Distressed pacing. Fidgeting	Rigid. Fists clenched. Knees pulled up. Pulling or pushing away. Striking out
Consolability	No need to console	Distracted or reassured by voice or touch	Unable to console, distract or reassure

From Warden et al. [51]. Reprinted with permission from Elsevier Limited

Managing Symptoms

Studies have shown that elderly cardiovascular patients are often either under-treated, or over-treated with burdensome interventions or polypharmacy. Before adding a new medication to manage symptoms at the EOL, it is imperative to examine the current prescriptions and consider removing non-essential medications that may be causing or aggravating symptoms. Symptom-alleviating medications such as diuretics, nitrates, and opioids may be favored over ACE inhibitors and beta-blockers. If a medication is to be added, the mantra of "start low and go slow, but get there" should be followed, meaning that low doses and slow titrations are employed while meticulously pursuing relief of symptoms [52].

In addition to pain and dyspnea, behavioral and psychological symptoms such as agitation and restlessness are common in CI patients at EOL [53]. Non-pharmacological interventions should be attempted as first line, including: reality orientation, music therapy, environment modification, and cognitive behavioral therapy. Neuroleptic medications may be required as second line, and these have been shown to be efficacious in short-term studies <6–12 weeks; however, there does not appear to be a sustained benefit >12 weeks and there is a risk of increased drowsiness and worsened quality of life [54]. Serotonin reuptake inhibitors may be used to treat depression, which is co-prevalent with frailty in many CVD patients.

Nutrition

Frail individuals may benefit from a high-caloric protein-rich diet. In a pilot trial, 29 patients with severe heart failure and non-edema weight loss >7.5 % were randomized to placebo or oral supplements containing 600 kcal and 20 g of protein; the supplements were associated with a 2.0 kg weight gain at 6 weeks and more importantly an improvement in quality of life as measured by the Minnesota Living with Heart Failure Questionnaire [55]. Testosterone supplements have been shown to increase muscle mass and functional capacity in patients with heart failure [56]. For those with CI and inability to swallow, a percutaneous endoscopic gastrostomy (PEG) may be considered in selected cases, although this is a complex issues which has been reviewed in detail elsewhere [57]. PEG placement to increase caloric and protein intake in frail patients with advanced CVD and CI would seem reasonable in theory, although the global prognosis and likelihood of meaningful functional improvement is not often favorable in such patients and there is no evidence to support this practice. PEG placement to prevent aspiration pneumonia is also not supported by the evidence.

Advanced Care Planning

The finding of frailty foreshadows a heightened risk and gravity of future cardiovascular decompensations. This is a cue to introduce the process of advanced directives and engage the patient in discussing their preferences for health care in the event that they become physically or mentally unable to speak for themselves and make their wishes known. Advanced directives may be elaborated in the living will or through the durable power of attorney for health care. The clinician should provide information about the prognosis, possibility of decompensation, therapeutic options that may be available at the various stages, and in turn, elicit the patient's preferences and values. The expected outcomes after a decompensation and/or invasive therapy should be framed around the individual patient and take into account their frailty, since frail patients have decreased resiliency and are less likely to recover favorably.

When making advanced decisions, dementia and dependency in ADLs are major risk factors for death or vegetative state after cardiopulmonary resuscitation (CPR), with dependency conferring a ≤3.5 % likelihood of survival to hospital discharge [58]. Eighty-four percent of surveyed physicians felt that a *Do Not Resuscitate* (DNR) order should be brought forth in patients with severe dementia [59], although this should not be extrapolated to delivering sub-standard care for the patient's other needs [60]. Partial DNR orders are frequently encountered in day-to-day practice, namely: "slow code", "chemical code" (drugs but no CPR), and menu-like checklists of options for arrest and pre-arrest maneuvers. For frail or CI patients, partial DNR orders may falsely appear to be a middle ground that appeases the patient and family while setting some limits of care, yet partial DNR orders carry the risk of causing increased confusion and harm to the patient [61].

Family participation is highly recommended in all advanced decisions, and becomes critical if the patient has CI. Ideally, advanced planning should be initiated before CI reaches severe stage to maximize the patient's first-hand input. For families of patients with severe CI, the surrogate decision making process varies according to geo-cultural factors, education level, religious beliefs, and understanding of disease and prognosis. Medical ethics consultation may be considered on a case-by-case basis.

Decision Making Capacity

Questions surrounding competency to make medical decisions often arise when caring for elderly patients with suspected CI. The MMSE can be used as an initial screening tool, whereby a score ≥24 suggests that the patient is competent, a score ≤17 suggests that the patient is not competent, and a score 18–23 is uncertain and merits further specialized assessment [62]. However, the MMSE has limitations; it lacks any direct assessment of executive function, which is a key element of competency. Problems with executive function can go unnoticed, particularly in a previously high-functioning patient who has preserved attention and language skills [63]. Neuropsychological expertise from a psychiatric consultant (or other service depending on the hospital) should be requested when there is substantial uncertainty about decision-making capacity.

Shared Decision Making

Shared decision making is the process by which clinicians and patients share information and work toward decisions about treatment chosen from medically reasonable options that are aligned with the patients' preferences, goals, and values. The patients' preferences are not always aligned with or known to the physicians delivering care. The landmark SUPPORT study showed that 47 % of physicians were aware of their patients' preference to avoid CPR, and 46 % of DNR orders were written ≤2 days before death [64]. Approximately 50 % of patients were in favor of trading increased survival time for increased quality-of-life [65]. When deciding about a proposed treatment, patients placed greater importance on the risk of cognitive or functional impairment than the risk of dying or the treatment burden [66]. When asked about CPR, many patients and family members believed that the success rate was >50 %, when in reality it was <15 % [61]. The Regional Study of Care for the Dying showed that half of patients with end-stage CVD were unaware that their risk of short-term mortality was high [67].

The aforementioned highlight the communication gaps that exist between physicians and patients. The presence of CI, depression, and anxiety further widen these gaps and render the shared decision making process even more challenging. A *Scientific Statement from the American Heart Association for Decision Making in Advanced Heart Failure* is available to guide clinicians through this vital process [68]. The focus of shared decision making should revolve around patient-centered outcomes rather than mortality in isolation.

Device Deactivation

One out of five patients with an implantable cardiac defibrillator (ICD) will receive shock(s) in the final days or weeks of life, and this is undeniably associated with discomfort and anxiety. Ninety percent of hospice facilities do not have device deactivation policies and at least one-third of patients are unaware that their ICD could be deactivated [69]; presumably this proportion is even higher in CI patients. Based on the ethical principal of *respect for autonomy*, withdrawal of life sustaining therapies such as pacemakers, ICDs, and ventricular assist devices should be explained as an option at the EOL. Ideally, if a patient is frail or CI, this should be discussed before implanting a device.

The Heart Rhythm Society has outlined the following action steps [70]: (1) confirm the patient's decision making capacity or identify an appropriate surrogate, (2) document the patient's wishes and discussion with regard to the specific device features to be deactivated, (3) ask an electrophysiologist or trained allied health professional to perform the deactivation. American guidelines do not differentiate between ICD or pacemaker deactivation, but they do acknowledge that the latter may be morally disturbing for some clinicians and prohibited by law in certain European countries.

Palliative and Hospice Care

Beyond the general benefits of PC in end-stage CVD [71], the services offered by PC teams are particularly well aligned to the special needs of frail and CI patients. When patients are judged to be too frail for a cardiac intervention, this is a resonating cue to think about transitioning to PC. This approach is visible in transcatheter heart valve clinics, where frailty is systematically measured and utilized by the multidisciplinary heart team to decide if an aortic valve replacement intervention is likely to be futile [72], in which case a PC strategy is to be recommended. The Transcatheter Heart Valve team at St. Paul's Hospital has outlined their tailored approach to incorporate PC in practice [73]. Although many patients are ill prepared for PC and tend to confuse PC and hospice care, they ultimately derive great benefits from ameliorated symptom control, emotional support, communication of important information, navigating the system, and transitioning to comfort/hospice care [74].

There are disconcerting trends towards more burdensome interventions and less palliative and hospice care in frail and CI patients with end-stage CVD. In the final year of life, advanced heart failure patients spend on average 30–40 days in hospital [75]. Many of these patients indicate a preference to die at home, although this is only achieved in a minority of cases. Those with CI are more likely to be transferred to die at the hospital because caretakers are not equipped to deal with their EOL symptoms at home. This is associated with patient and family dissatisfaction and enormous strain on the health care system. The annualized cost of inpatient care for patients with profound dementia has been estimated at $7.3–9.0 billion, and $3.8 billion for frail patients with ≤ 1 % probability of surviving 2 months [76]. There may be geo-cultural variability in the level of care for patients with dementia, as an American study found that dementia was an independent predictor of increased Medicare costs for heart failure patients in the last 6 months of life [77], whereas a Canadian study found the opposite [78].

Conclusion

Ensuring a *good death* for our CVD patients with CI or frailty may be viewed through different lenses: the patient's, and secondarily, the family's, the clinician's, and the society's. Research tools exist to measure quality of dying for the patient and family [79], and guidelines exist to enlighten best practices for the clinician [80], but at a fundamental level there are three basic tenets: (1) provide meticulous relief of physical symptoms, (2) provide compassionate holistic care that addresses spiritual and psychological needs, and (3) engage all parties in shared decision making and advanced care planning [81]. These tenets, although applicable to all, are especially vital in the care of frail and cognitively impaired patients with advanced CVD.

References

1. Afilalo J, Alexander KP, Mack MJ, Maurer MS, Green P, Allen LA, et al. Frailty assessment in the cardiovascular care of older adults. J Am Coll Cardiol. 2014;63(8):747–62.
2. Dodson JA, Truong T-TN, Towle VR, Kerins G, Chaudhry SI. Cognitive impairment in older adults with heart failure: prevalence, documentation, and impact on outcomes. Am J Med. 2013;126(2):120–6.
3. Huijts M, van Oostenbrugge RJ, Duits A, Burkard T, Muzzarelli S, Maeder MT, et al. Cognitive impairment in heart failure: results from the Trial of Intensified versus standard Medical therapy in Elderly patients with Congestive Heart Failure (TIME-CHF) randomized trial. Eur J Heart Fail. 2013;15(6):699–707.
4. Vogels RLC, Scheltens P, Schroeder-Tanka JM, Weinstein HC. Cognitive impairment in heart failure: a systematic review of the literature. Eur J Heart Fail. 2007;9(5):440–9.
5. Kindermann I, Fischer D, Karbach J, Link A, Walenta K, Barth C, et al. Cognitive function in patients with decompensated heart failure: the cognitive impairment in heart failure (CogImpair-HF) study. Eur J Heart Fail. 2012;14(4):404–13.
6. Collard RM, Boter H, Schoevers RA, Oude Voshaar RC. Prevalence of frailty in community-dwelling older persons: a systematic review. J Am Geriatr Soc. 2012;60(8):1487–92.
7. Fried LP, Tangen CM, Walston J, Newman AB, Hirsch C, Gottdiener J, et al. Frailty in older adults: evidence for a phenotype. J Gerontol A Biol Sci Med Sci. 2001;56(3):M146–56.
8. Rockwood K, Song X, MacKnight C, Bergman H, Hogan DB, McDowell I, et al. A global clinical measure of fitness and frailty in elderly people. Can Med Assoc J. 2005;173(5): 489–95.
9. Shamliyan T, Talley KMC, Ramakrishnan R, Kane RL. Association of frailty with survival: a systematic literature review. Ageing Res Rev. 2012;12.
10. Afilalo J, Karunananthan S, Eisenberg MJ, Alexander KP, Bergman H. Role of frailty in patients with cardiovascular disease. Am J Cardiol. 2009;103(11):1616–21.
11. Fried LP, Ferrucci L, Darer J, Williamson JD, Anderson G. Untangling the concepts of disability, frailty, and comorbidity: implications for improved targeting and care. J Gerontol A Biol Sci Med Sci. 2004;59(3):255–63.
12. Katz S, Ford AB, Moskowitz RW, Jackson BA, Jaffe MW. Studies of illness in the aged. The index of ADL: a standardized measure of biological and psychosocial function. JAMA. 1963;185:914–9.
13. Mahoney FI, Barthel DW. Functional evaluation: the Barthel index. Md State Med J. 1965; 14:61–5.
14. Fillenbaum GG, Smyer MA. The development, validity, and reliability of the OARS multidimensional functional assessment questionnaire. J Gerontol. 1981;36(4):428–34.
15. Walter LC, Brand RJ, Counsell SR, Palmer RM, Landefeld CS, Fortinsky RH, et al. Development and validation of a prognostic index for 1-year mortality in older adults after hospitalization. JAMA. 2001;285(23):2987–94.
16. Frohnhofen H, Hagen O, Heuer HC, Falkenhahn C, Willschrei P, Nehen HG. The terminal phase of life as a team-based clinical global judgment: prevalence and associations in an acute geriatric unit. Zeitschrift fur Gerontologie und Geriatrie. 2011;44(5):329–35.

17. Afilalo J, Eisenberg MJ, Morin J-F, Bergman H, Monette J, Noiseux N, et al. Gait speed as an incremental predictor of mortality and major morbidity in elderly patients undergoing cardiac surgery. J Am Coll Cardiol. 2010;56(20):1668–76.
18. Molloy DW, Standish TI. A guide to the standardized Mini-Mental State Examination. Int Psychogeriatr. 1997;9 Suppl 1:87–94; discussion 143–50.
19. Nasreddine ZS, Phillips NA, Bédirian V, Charbonneau S, Whitehead V, Collin I, et al. The Montreal Cognitive Assessment, MoCA: a brief screening tool for mild cognitive impairment. J Am Geriatr Soc. 2005;53(4):695–9.
20. Cameron J, Worrall-Carter L, Page K, Stewart S, Ski CF. Screening for mild cognitive impairment in patients with heart failure: montreal cognitive assessment versus mini mental state exam. Eur J Cardiovasc Nurs. 2013;12(3):252–60.
21. Davis KK, Allen JK. Identifying cognitive impairment in heart failure: a review of screening measures. Heart Lung. 2013;42(2):92–7.
22. Hauptman PJ, Swindle J, Hussain Z, Biener L, Burroughs TE. Physician attitudes toward end-stage heart failure: a national survey. Am J Med. 2008;121(2):127–35.
23. Coventry PA, Grande GE, Richards DA, Todd CJ. Prediction of appropriate timing of palliative care for older adults with non-malignant life-threatening disease: a systematic review. Age Ageing. 2005;34(3):218–27.
24. Peterson PN, Rumsfeld JS, Liang L, Albert NM, Hernandez AF, Peterson ED, et al. A validated risk score for in-hospital mortality in patients with heart failure from the American Heart Association get with the guidelines program. Circ Cardiovasc Qual Outcomes. 2010;3(1): 25–32.
25. Lee DS, Austin PC, Rouleau JL, Liu PP, Naimark D, Tu JV. Predicting mortality among patients hospitalized for heart failure: derivation and validation of a clinical model. JAMA. 2003;290(19):2581–7.
26. Levy WC, Mozaffarian D, Linker DT, Sutradhar SC, Anker SD, Cropp AB, et al. The Seattle Heart Failure Model: prediction of survival in heart failure. Circulation. 2006;113(11): 1424–33.
27. Huynh BC, Rovner A, Rich MW. Long-term survival in elderly patients hospitalized for heart failure: 14-year follow-up from a prospective randomized trial. Arch Intern Med. 2006;166(17):1892–8.
28. Rozzini R, Sabatini T, Frisoni GB, Trabucchi M. Frailty is a strong modulator of heart failure-associated mortality. Arch Intern Med. 2003;163(6):737–8; author reply 738.
29. Pilotto A, Addante F, Franceschi M, Leandro G, Rengo G, D'Ambrosio P, et al. Multidimensional prognostic index based on a comprehensive geriatric assessment predicts short-term mortality in older patients with heart failure. Circ Heart Fail. 2010;3(1):14–20.
30. Afilalo J, Mottillo S, Eisenberg MJ, Alexander KP, Noiseux N, Perrault LP, et al. Addition of frailty and disability to cardiac surgery risk scores identifies elderly patients at high risk of mortality or major morbidity. Circ Cardiovasc Qual Outcomes. 2012;5(2):222–8.
31. Purser JL, Kuchibhatla MN, Fillenbaum GG, Harding T, Peterson ED, Alexander KP. Identifying frailty in hospitalized older adults with significant coronary artery disease. J Am Geriatr Soc. 2006;54(11):1674–81.
32. Gharacholou SM, Roger VL, Lennon RJ, Rihal CS, Sloan JA, Spertus JA, et al. Comparison of frail patients versus nonfrail patients' ≥65 years of age undergoing percutaneous coronary intervention. Am J Cardiol. 2012;109:1569–75.
33. Adler ED, Goldfinger JZ, Kalman J, Park ME, Meier DE. Palliative care in the treatment of advanced heart failure. Circulation. 2009;120(25):2597–606.
34. Goodman C, Evans C, Wilcock J, Froggatt K, Drennan V, Sampson E, et al. End of life care for community dwelling older people with dementia: an integrated review. Int J Geriatr Psychiatry. 2010;25(4):329–37.
35. Schulz R, Mendelsohn AB, Haley WE, Mahoney D, Allen RS, Zhang S, et al. End-of-life care and the effects of bereavement on family caregivers of persons with dementia. N Engl J Med. 2003;349(20):1936–42.

36. Setoguchi S, Glynn RJ, Stedman M, Flavell CM, Levin R, Stevenson LW. Hospice, opiates, and acute care service use among the elderly before death from heart failure or cancer. Am Heart J. 2010;160(1):139–44.
37. Bekelman DB, Rumsfeld JS, Havranek EP, Yamashita TE, Hutt E, Gottlieb SH, et al. Symptom burden, depression, and spiritual well-being: a comparison of heart failure and advanced cancer patients. J Gen Intern Med. 2009;24(5):592–8.
38. Lunney JR, Lynn J, Foley DJ, Lipson S, Guralnik JM. Patterns of functional decline at the end of life. JAMA. 2003;289(18):2387–92.
39. Gill TM, Gahbauer EA, Han L, Allore HG. Trajectories of disability in the last year of life. N Engl J Med. 2010;362(13):1173–80.
40. Alosco ML, Spitznagel MB, Cohen R, Raz N, Sweet LH, Josephson R, et al. Reduced cerebral perfusion predicts greater depressive symptoms and cognitive dysfunction at a 1-year follow-up in patients with heart failure. Int J Geriatr Psychiatry. 2014;29(4):428–36.
41. Vogels RLC, Oosterman JM, Laman DM, Gouw AA, Schroeder-Tanka JM, Scheltens P, et al. Transcranial Doppler blood flow assessment in patients with mild heart failure: correlates with neuroimaging and cognitive performance. Congest Heart Fail. 2008;14(2):61–5.
42. Zuccala G, Onder G, Pedone C, Carosella L, Pahor M, Bernabei R, et al. Hypotension and cognitive impairment: selective association in patients with heart failure. Neurology. 2001; 57(11):1986–92.
43. Zuccalà G, Marzetti E, Cesari M, Monaco Lo MR, Antonica L, Cocchi A, et al. Correlates of cognitive impairment among patients with heart failure: results of a multicenter survey. AJM. 2005;118(5):496–502.
44. Criner GJ. Ambulatory home oxygen: what is the evidence for benefit, and who does it help? Respir Care. 2013;58(1):48–64.
45. Kielb SA, Ancoli-Israel S, Rebok GW, Spira AP. Cognition in obstructive sleep apnea-hypopnea syndrome (OSAS): current clinical knowledge and the impact of treatment. Neuromol Med (Humana Press Inc). 2012;14(3):180–93.
46. Almeida OP, Garrido GJ, Beer C, Lautenschlager NT, Arnolda L, Flicker L. Cognitive and brain changes associated with ischaemic heart disease and heart failure. Eur Heart J. 2012;33(14): 1769–76.
47. Rolland Y, Abellan van Kan G, Gillette-Guyonnet S, Vellas B. Cachexia versus sarcopenia. Curr Opin Clin Nutr Metab Care. 2011;14(1):15–21.
48. Wolfe RR. Optimal nutrition, exercise, and hormonal therapy promote muscle anabolism in the elderly. J Am Coll Surg. 2006;202(1):176–80.
49. Liu CK, Fielding RA. Exercise as an intervention for frailty. Clin Geriatr Med. 2011;27(1): 101–10.
50. Theou O, Stathokostas L, Roland KP, Jakobi JM, Patterson C, Vandervoort AA, et al. The effectiveness of exercise interventions for the management of frailty: a systematic review. J Aging Res. 2011;2011:569194.
51. Warden V, Hurley AC, Volicer L. Development and psychometric evaluation of the Pain Assessment in Advanced Dementia (PAINAD) scale. J Am Med Dir Assoc. 2003;4(1):9–15.
52. Koller K, Rockwood K. Frailty in older adults: implications for end-of-life care. Cleve Clin J Med. 2013;80(3):168–74.
53. van der Steen JT, Gijsberts MJ, Knol DL, Deliens L, Muller MT. Ratings of symptoms and comfort in dementia patients at the end of life: comparison of nurses and families. Palliat Med. 2009;23(4):317–24.
54. Robinson L, Hughes J, Daley S. End-of-life care and dementia. Rev Clin Gerontol. 2005.
55. Rozentryt P, von Haehling S, Lainscak M, Nowak JU, Kalantar-Zadeh K, Polonski L, et al. The effects of a high-caloric protein-rich oral nutritional supplement in patients with chronic heart failure and cachexia on quality of life, body composition, and inflammation markers: a randomized, double-blind pilot study. J Cachex Sarcopenia Muscle. 2010;1(1):35–42.
56. Toma M, McAlister FA, Coglianese EE, Vidi V, Vasaiwala S, Bakal JA, et al. Testosterone supplementation in heart failure: a meta-analysis. Circ Heart Fail. 2012;5(3):315–21.

57. Sampson EL, editor. Enteral tube feeding for older people with advanced dementia. Cochrane Database Syst Rev. 2009;(2):CD007209.
58. Ebell MH, Jang W, Shen Y, Geocadin RG, Get With the Guidelines–Resuscitation Investigators. Development and validation of the Good Outcome Following Attempted Resuscitation (GO-FAR) score to predict neurologically intact survival after in-hospital cardiopulmonary resuscitation. JAMA Intern Med. 2013;173(20):1872–8.
59. Kelley AS, Reid MC, Miller DH, Fins JJ, Lachs MS. Implantable cardioverter-defibrillator deactivation at the end of life: a physician survey. Am Heart J. 2009;157(4):702–8.e1.
60. Chen JLT, Sosnov J, Lessard D, Goldberg RJ. Impact of do-not-resuscitation orders on quality of care performance measures in patients hospitalized with acute heart failure. Am Heart J. 2008;156(1):78–84.
61. Sanders A, Schepp M, Baird M. Partial do-not-resuscitate orders: a hazard to patient safety and clinical outcomes? Crit Care Med. 2011;39(1):14–8.
62. Karlawish JHT, Casarett DJ, James BD, Xie SX, Kim SYH. The ability of persons with Alzheimer disease (AD) to make a decision about taking an AD treatment. Neurology. 2005; 64(9):1514–9.
63. Gaviria M, Pliskin N, Kney A. Cognitive impairment in patients with advanced heart failure and its implications on decision-making capacity. Congest Heart Fail. 2011;17(4):175–9.
64. A controlled trial to improve care for seriously ill hospitalized patients. The study to understand prognoses and preferences for outcomes and risks of treatments (SUPPORT). The SUPPORT Principal Investigators. JAMA. 1995;274(20):1591–8.
65. Dev S, Abernethy AP, Rogers JG, O'Connor CM. Preferences of people with advanced heart failure-a structured narrative literature review to inform decision making in the palliative care setting. Am Heart J. 2012;164(3):313–5.
66. Fried TR, Bradley EH, Towle VR, Allore H. Understanding the treatment preferences of seriously ill patients. N Engl J Med. 2002;346(14):1061–6.
67. McCarthy M, Hall JA, Ley M. Communication and choice in dying from heart disease. J R Soc Med. 1997;90(3):128–31.
68. Allen LA, Stevenson LW, Grady KL, Goldstein NE, Matlock DD, Arnold RM, et al. Decision making in advanced heart failure: a scientific statement from the American Heart Association. Circulation. 2012;125(15):1928–52.
69. Pedersen SS, Chaitsing R, Szili-Torok T, Jordaens L, Theuns DAMJ. Patients' perspective on deactivation of the implantable cardioverter-defibrillator near the end of life. Am J Cardiol. 2013;111(10):1443–7.
70. Lampert R, Hayes DL, Annas GJ, Farley MA, Goldstein NE, Hamilton RM, et al. HRS expert consensus statement on the management of cardiovascular implantable electronic devices (CIEDs) in patients nearing end of life or requesting withdrawal of therapy. Heart Rhythm. 2010;7(7):1008–26.
71. Low J, Pattenden J, Candy B, Beattie JM, Jones L. Palliative care in advanced heart failure: an international review of the perspectives of recipients and health professionals on care provision. J Card Fail. 2011;17(3):231–52.
72. Lindman B, Alexander KP, O'Gara PT, Afilalo J. Futility, benefit, and transcatheter aortic valve replacement. JACC Cardiovasc Interv. 2014.
73. Lauck S, Garland E, Achtem L, Forman J, Baumbusch J, Boone R, et al. Integrating a palliative approach in a transcatheter heart valve program: bridging innovations in the management of severe aortic stenosis and best end-of-life practice. Eur J Cardiovasc Nurs. 2014;13(2): 177–84.
74. Metzger M, Norton SA, Quinn JR, Gramling R. Patient and family members' perceptions of palliative care in heart failure. Heart Lung. 2013;42(2):112–9.
75. Reed SD, Li Y, Dunlap ME, Kraus WE, Samsa GP, Schulman KA, et al. In-hospital resource use and medical costs in the last year of life by mode of death (from the HF-ACTION randomized controlled trial). Am J Cardiol. 2012;110(8):1150–5.
76. Zilberberg MD, Shorr AF. Economics at the end of life: hospital and ICU perspectives. Semin Respir Crit Care Med. 2012;33(4):362–9.

77. Unroe KT, Greiner MA, Hernandez AF, Whellan DJ, Kaul P, Schulman KA, et al. Resource use in the last 6 months of life among medicare beneficiaries with heart failure, 2000–2007. Arch Intern Med. 2011;171(3):196–203.
78. Kaul P, McAlister FA, Ezekowitz JA, Bakal JA, Curtis LH, Quan H, et al. Resource use in the last 6 months of life among patients with heart failure in Canada. Arch Intern Med. 2011;171(3):211–7.
79. van Soest-Poortvliet MC, van der Steen JT, Zimmerman S, Cohen LW, Munn J, Achterberg WP, et al. Measuring the quality of dying and quality of care when dying in long-term care settings: a qualitative content analysis of available instruments. J Pain Symptom Manage. 2011;42(6):852–63.
80. Gove D, Sparr S, Dos Santos Bernardo AMC, Cosgrave MP, Jansen S, Martensson B, et al. Recommendations on end-of-life care for people with dementia. J Nutr Health Aging. 2010;14(2):136–9.
81. Lawrence V, Samsi K, Murray J, Harari D, Banerjee S. Dying well with dementia: qualitative examination of end-of-life care. Br J Psychiatry. 2011;199(5):417–22.

Index

S.J. Goodlin, M.W. Rich (eds.), *End-of-Life Care in Cardiovascular Disease*,
DOI 10.1007/978-1-4471-6521-7

MIX
Papier aus verantwortungsvollen Quellen
Paper from responsible sources
FSC® C105338

If you have any concerns about our products,
you can contact us on
ProductSafety@springernature.com

In case Publisher is established outside the EU,
the EU authorized representative is:
Springer Nature Customer Service Center GmbH
Europaplatz 3, 69115 Heidelberg, Germany

Printed by Libri Plureos GmbH
in Hamburg, Germany